MORE SNOWDROPS

Published by Snowdrop Editions
Shropshire

ISBN 978-1-84396-691-3

This work has been registered
with the British Library and the American
Library of Congress

Front cover illustration
Galanthus 'Ernie Cavallo'

Typesetting and pre-press production
eBook Versions
27 Old Gloucester Street
London WC1N 3AX
www.ebookversions.com

For Ernie and Susan with love,
and in memory of Joan, Michael, Sue
and Lady Carolyn Elwes

G. 'Carolyn Elwes'

MORE SNOWDROPS

Freda Cox

SNOWDROP EDITIONS

Contents

ACKNOWLEDGMENTS

A massive 'Thank you' to Joseph Hawthorne for all of his help and hard work towards compiling this latest book. Also all the helpful Galanthophiles out there who have provided photographs and information. We wouldn't have got there without you and your newly named snowdrops.

Preface

This will be my final update of newly named snowdrops. Twelve months ago it seemed a good idea to research the latest Galanthus names and descriptions for a small updated list to send to like-minded friends.

Thinking there would be roughly around 300 new snowdrops, I found there are now well over 700 – though I am sure there are still more to discover, draw and describe. New snowdrops and names are constantly being introduced. Surely there has to be an end somewhere?

This project turned out to be far larger than anticipated, necessitating a new volume to add to the first and second editions of my previous work, *A Gardener's Guide to Snowdrops*, published by Crowood Press, which also contains technical and horticultural information.

Some snowdrops have all the details one could wish for, while others barely get a mention – apart from a picture and/or name. The whole thing isn't helped either by so many anomalies amongst snowdrops, which muddy the water. Many names appear to be very transient: one minute they are there, the next they have disappeared along with their snowdrop. There are descriptions but no illustrations, and illustrations without descriptions. Not to mention all the other anomalies of which here are just a few:

A number of snowdrops can show colour changes year on year. For instance, snowdrops with yellow markings suddenly produce green markings. The pure white albino snowdrop can produce green marks. 'Normal' snowdrops go aberrant. Galanthus 'Bright Eyes' appeared with four by four segments when it is normally only three by three. Not the only snowdrop to suddenly do that. *G.* Atkinsii appears with four outer segments and a mess of inner segments.

G. 'Altered Image' originated from a bulb with four by four flowers but can produce three by three, four by four or five by five flowers. *G.* 'Mrs Thompson' produces three heads on one scape, although known for her highly erratic behaviour. *G.* 'Phantom' can produce two very different looking flowers from the same bulb when established. *G.* 'Ecusson d'Or' appears to have split into two types, short and long pedicels. *G.* 'Margaret Biddulph' with bright green-yellow marks, and a twin *G.* 'Wol Staines'.

I have used the spelling of names as they appeared with photographs and information received, even when I have thought this possibly suspect. And so it goes on – anomaly after anomaly to trap the unwary!

Despite all the different colour changes, marks and shapes I still think nothing comes close to the tiny *G. nivalis* so eagerly awaited to herald in the Spring and which brightens long, grey winter days.

But beware – once Galanthomania takes hold, it doesn't let go. Every year I say enough is enough and yet, when new names start appearing out comes the paintbrush and I am off again. This really is the very last time!

Freda Cox
Shropshire, 2022

Naming Snowdrops

It is of paramount importance that a new snowdrop proves its worth before being named. It must be completely new and not one already available under another name. It must be reliable and distinct, continue to come true each year in different situations, and most importantly, be different and worthy enough to warrant a name.

The naming of plants is strictly regulated by the International Code of Botanical Nomenclature for Wild Plants and the International Code of Nomenclature for Cultivated Plants, also known as the Cultivated Plant Code. These set out rules, recommendations and guidelines for naming and registering new plants which must be strictly adhered to. The ICNCP has also established a system of recording permanent reference specimens of the plant type. These can either be pressed herbarium specimens or meticulously accurate illustrations.

It is sensible to register newly named snowdrop cultivars with the Royal General Bulb Growers Association, Hillegom, Netherlands, established in 1860.

The KAVB, (Koninkljke Algemeene Vereenniging voor Bloembollencultuur) maintains a database register of bulb cultivars that can be accessed via their website.

ICRA (International Cultivar Registration Authority) was set up in an attempt to regulate and simplify the naming of plant cultivars and prevent the duplicate use of cultivar and group epithets. Data is updated every four years. It is a voluntary international system and registering a cultivar name does not protect that name or infer any legal right.

These are dealt with under the National Plant Breeders Rights or Plant Patents. ICRA will check the name is not already in use in that specific genera of plants, and that is has not been used previously. It is important that a full description of the plant is submitted with the new name, including the plant's parentage, etc.

The International Union for the Protection of New Varieties of Plants offers legal protection to plant breeders when introducing and registering new cultivars.

The Royal Horticultural Society's joint Rock Garden Plant Committee makes awards for outstanding plants and will offer advice on new snowdrops.

Any new name must consist of the plant's scientific botanical name, in this case *Galanthus*, followed by a unique epithet that is generally a vernacular term enclosed in single quotation marks. Before 1959 Latin names were often also used for cultivar epithets, leading to great confusion when they were misrepresented as botanical names. After 1st January 1959 the particular epithet had to be in a modern language.

If the new name complies with all the correct criteria, it then has to be printed in a legitimate, widely available publication before it becomes accepted.

There is still much room for improvement in snowdrop naming and listing. This is where establishing and recording details of a basic 'type' for each snowdrop, including an illustration, has great advantages for reference. Many old snowdrops have received new names in an effort to increase clarity and conformity among cultivar lists and names.

If you are sure your new snowdrop fills all the correct criteria then, and only then, is the time to think of a name.

Snowdrop Immortals

Snowdrop Immortals, initiated by Alan Street and based on the French Immortals, which was an association of like-minded people who met each year, dedicated to keeping the French language pure.

To qualify as a Snowdrop Immortal you must have a snowdrop named after you, using both your first name and surname. Immortals include:

Ruby Baker, Bill Boardman, June Boardman, David Bromley, Ernie Cavallo, Ray Cobb, Veronica Cross, Brian Ellis, Carolyn Elwes, Yvonne Hay, Bryan Hewitt, Jörg Lebsa, Dorothy Lucking, Robert Mackenzie, Brian Mathew, Sally Pasmore, Celia Sawyer and Audry Vockins.

New Species 2022

Galanthus Samothracicus

Kit Tan, Burkhard Biel and Sonja Siljak-Yakovlev,(2014) *Galanthus samothracicus* (Amaryllidaceae) from the island of Samothraki, north-eastern Greece. Phytologia Balcanica, 20(1): 65 – 70.

Galanthus samothracicus has now been established as a new and separate *Galanthus* species. On 11th April 2006, Burkhard Biel took a botanical expedition to the Aegean Island of Samothraki, north-eastern Greece. Here he discovered seed-heads on groups of snowdrops. Returning to the island in 2009 and 2011 further snowdrops were discovered in other locations on the island. *G. nivalis* is not found in the Aegean and although related to *G. nivalis*, the new snowdrop was considered a local endemic on the island. Biel collected plants which he grew on in his garden in Germany, sending others to Copenhagen Botanic Garden.

It was ascertained that although very similar to standard *G. nivalis* there are small, but distinct morphological differences including segment markings and leaves in the new snowdrops, named *G. samothracicus*, as well as smaller amounts of DNA in the cell nucleus (genome).

Leaves are applanate with flat to slightly subrevolute margins, without a paler central median line, smooth and shiny green, losing the typical glaucous sheen as they mature. There is only a single broad or 'U'- shaped green mark above the sinus.

Plants were also found to be extremely frost hardy surviving temperatures down to -20°C. Flowers are slender and rounded. Leaves erect to arching, slender, broadening on upper third, blue-green but distinctively greener at maturity, losing their glaucous sheen. Outer segments long, slender, rounded, lightly longitudinally ridged, slender claw, tapering to bluntly pointed apex. Inner segments a broad or inverted 'U'- shaped mid-green mark above sinus, paling on basal side. Light perfume. Flowering January-February in Greece and mid-December to late February in cultivation, *G. samothracicus* grows in humus rich soil in damp and seasonally wet habitats near streams and springs from between 5-80 m. Height 16-31cm.

Galanthus bursanus
as described by. Yildiz Konca, Dimitri A. Zubov and Aaron P. Davis 2019

In 2014, Yildiz Konca and Dimitri A. Zubov were researching the Marmara sea region of north-west Turkey near the City of Bursa in Bursa Province. They discovered two small populations of a snowdrop they subsequently named as a new species. *G. bursanus*, after the region in which it was found. This species is autumn flowering usually from September, with a very small distribution area of around only 8 square km. 500m above sea level on the wooded Korucak Dagi ridge. The snowdrop grows beneath Turkey Oak, *Quercus cerris*, on rocky limestone outcrops.

The species was first described in 2019. Other species in the region include *G, gracilis*; *G. plicatus* subs *plicatus*; *G. plicatus* subs *byzantinus*; *G. trojanus* and *G x valentinei* nothosubsp. *sublicatus* – all of which are different in respect of being winter/spring flowering between December and April, with leaves well developed at flowering.

The new snowdrop species was first discovered as two small populations with either no leaves at flowering or very short leaves. Despite resembling *G. reginae-olgae* subs. *reginae-olgae*, there were very distinct differences and after careful examination it was concluded the differences were such they did not resemble any other locally growing species. This snowdrop was then identified as a completely new species, *G. bursanus*.

Further field studies were carried out between 2016 and 2018, including detailed examination with microscopes and digital camera software.

The bulb is narrowly ovoid with whitish bulb scales, fully covered by a yellow-brown papery tunic, and having whitish adventitious roots. Leaves are applanate-narrowly explicative, strap-shaped, either absent or very short at flowering and slightly twisted when mature with two noticeable, short longitudinal folds. Leaf surfaces are dark green, blue-grey-glaucous with greener margins and a broad central glaucous stripe and small white tip. There are usually two scapes, rarely one or three, with a dark green-glaucous, papery spathe, equal in length to the pedicel at flowering. The ovary is globose to narrowly elliipsoid and bright green.

Flowers are narrowly ovoid to pear-shaped in bud and strongly fragrant, having three outer and three inner segments. Outer segments are narrowly ovoid and indistinctly longitudinally ridged, white, with a short claw and pointed apex. Inner segments are half the length of outers, white with an apical sinus notch and a 'comma' shaped green mark either side of the sinus, which can join, a second mark towards the base can merge to form a roughly 'X'- shaped mark across the segment. Six orange stamens are arranged in two whorls.

G. bursanus flowers between October and January in the wild and late September to December in cultivation.

Although there are similarities there are also very noticeable differences but it is considered *G bursanus* is possibly related to *G. plicatus* subs. *byzantinus* and *G. x valentinei* nothosubsp. *sublicatus*, and again possibly to *G. reginae-oilgae* subs. *reginae-olgae*.

The exact locations of *G. bursanus* are not generally revealed, to avoid risk of unlawful collecting – especially as the species is classed as critically endangered. Height 18-25cm.

Directory of Newly Added Snowdrops

'Adam'
Attractive, heavily green marked virescent, Inverse Poculiform flowers on erect sapes with rounded ovary. Leaves medium, splayed. grey-green, paler median line. Outer segments broad, flattened, rounded at apex. Wide mid-green mark above apex almost to base. Inner segments shorter than outer, green mark above sinus almost to base, narrow white margin. Named for the late, friend of John Massey, Ashwood Garden and Nursery, Kingswinford, Winter/Spring.

'Adamite'
G. reginae-olgae subs *reginae-olgae* with tall, strong erect scapes and oblong green ovary. Flowers have unusual pale yellow-green colouring across segments rather than the usual white. The colouring is strongest in bud and when plants are grown in shade but it is paler than in *G.* 'Uranium', and also earlier flowering, generally late September to mid-October. Leaves erect, applanate, slender, with colouring darker on upper surface, short at flowering. Outer segments, slender tapering to pointed apex, tinged green. Inner segments green mark above sinus. Joe Sharman, Monksilver Nursery, Cambridge.

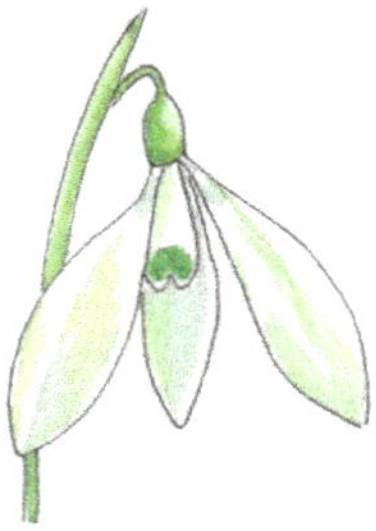

'Ade-Le'
Good greenmarked flowers. Outer segments broad, incurved, bluntly pointed at apex with merging green lines above apex to half of segment. Inner segments slightly flared, broad inverted mid-green 'V'-shaped mark above sinus, two paler merging ovals towards base. Hagen Engelmann, Garten in den Wiesen, Germany.

'Adriana'
Good green marks on well-shaped flowers. Outer segments rounded to bluntly pointed apex, longitudinally ridged, green mark above apex into ridges to half of segment. Inner segments green mark above sinus. Snowdropfevers.

'Aero'
Sturdy, shorter green-marked *G. nivalis*. Outer and inner segments green marks. Slovenia. 12cm.

'Alaina'
G. reginae-olgae subs. *vernalis* with neat looking, early, green-tipped flowers. Outer segments tapering to bluntly pointed apex, small green marks above apex. Inner segments green inverted 'V- shaped mark above sinus. Winter/Spring.

'Alan's Long Ovary'
As the name implies, flowers suspended from slender pedicel beneath long, slender, dark-green ovary. Leaves applanate, erect, slender, bluegreen. Outer segments rounded to bluntly pointed apex. Inner segments, narrow dark-green inverted 'V'-shaped mark above sinus. Winter/Spring. 16cm.

'Alban Arthan'
Rounded green ovary. Outer segments rounded, tapering to pointed apex. Inner segments green inverted broad 'U'-shaped mark above sinus. Named for the seasonal festival of the Winter Solstice. Winter.

'Albatross'
Good, short but very vigorous new *G. byzantinus* with well-rounded flowers and rounded green ovary. Leaves slender, splayed, green. Outer segments well rounded, tapering to bluntly pointed apex. Inner segments dark-green inverted 'V'-shaped mark with rounded ends above sinus and second dark-green mark towards base. Bulks up well and quickly. Andy Byfield. 2020. 8cm.

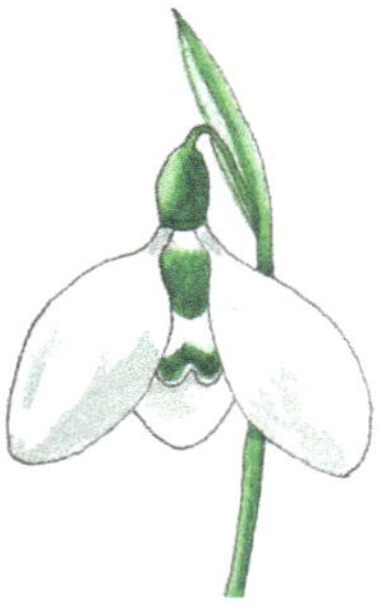

'Alexandra Beryl'
Autumn-flowering *G. reginae-olgae* subsp. *reginae-olgae*. Leaves applanate, short or absent at flowering. Outer segments rounded, lightly longitudinally ridged, tapering to pointed apex. Inner segments good heart-shaped mark above sinus. November. John Lonsdale, Edgewood Gardens, North East USA. 2021. 8cm.

'Alexandrite'
Virescent *G. ikariae* with slender-looking, green-marked flowers on erect scapes. Leaves supervolute, broad, erect to arching, green. Outer segments slender, incurved, bluntly pointed at apex, variable green lines above apex to half of segment. Inner segments good green mark above sinus to base, narrow white margin. Alan Street, Berkshire c. 1990s. February/March.

'Alice's Late'
Later-flowering *G. nivalis* with small, slender flowers. Leaves applanate, erect, slender, dark blue-green. Outer segments tapering to pointed apex. Inner segments, green inverted 'V'-shaped mark above sinus. A good Cornish variant flowering over a long period and invaluable for helping extend the season. Selected by Lady Alice Boyd, the Ince Estate, Cornwall and named by Beggars Roost Nursery. February-March. 14cm.

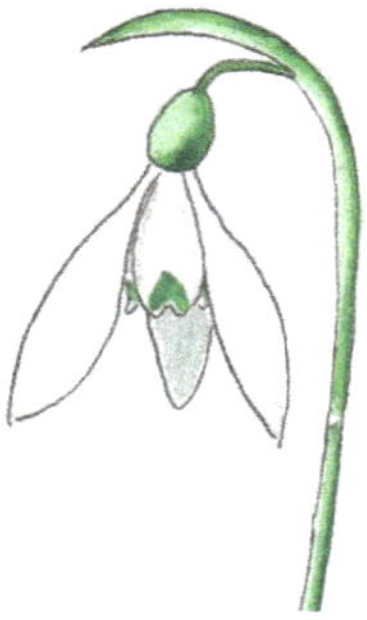

"Alles im Grunnen Bereich"
Attractive, well-shaped flowers with delicate green marked segments and slender, rounded ovary. Outer segments broad, tapering to pointed apex, light green wash from above apex towards base. Inner segments broad light-green mark above sinus diffusing towards base.

'Alpha Apricot Ice Cream'
Snowdrop with a faint blush of cream (apricot) on the inner segments and oblong, rounded green ovary. Outer segments broad, tapering to a pointed apex.Inner segments tinged apricot/cream with spreading green 'V'-shaped mark above sinus which can divide into two marks, and second green oval towards base. Ruslan Mishustin, Ukraine.

'Alpha Autumn Miracle'
Spreading green-marked flowers. Outer segments broad, flattened, rounded at apex, lightly longitudinally ridged, green mark above apex and second mark towards base. Inner segments spreading inverted green 'V'-shaped mark above sinus lightly joining second deep mark almost to base. Ruslan Mishustin, Ukraine. Autumn/Winter.

'Alpha Bermudian Triangle'
Rounded flowers beneath slender green ovary. Outer segments rounded, rounded at apex, slender, elongated green heart-shaped mark above apex. Inner segments single green mark above sinus towards base. Narrow white margin. Ruslan Mishustin, Ukraine.

'Alpha Boiling Milk'
Balloon-shaped, rounded flowers beneath small rounded green ovary. Outer segments broad, rounded, rounded at apex, longitudinally ridged. Inner segments, filled in inverted 'U'-shaped mark above sinus. Ruslan Mishustin, Ukraine.

'Alpha Bonus'
Autumn-flowering snowdrop with large, distinctively spoon-shaped outer segments, lightly longitudinally ridged, Inner segments, slender green mark above sinus with squared ends and second mark fading towards base. October Ruslan Mishustin, Ukraine. 2021. 20cm.

'Alpha Boo'
Erect scapes with slender green ovary and somewhat pointed flowers. Outer segments rounded tapering to pointed apex, lightly textured. Inner segments small green mark above sinus. Ruslan Mishustin, Ukraine.

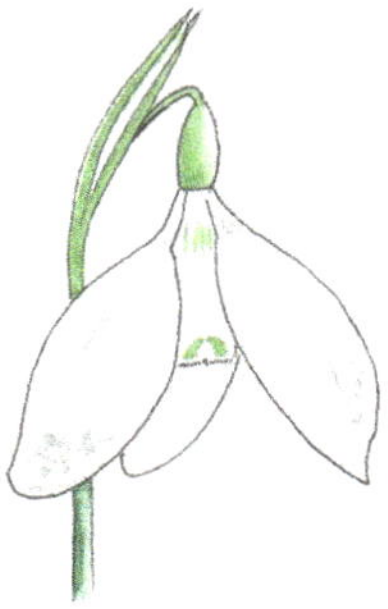

'Alpha Cat's Toy'
Autumn-flowering with slender ovary above rounded flowers. Outer segments tapering to pointed apex. Inner segments 'V'-shaped green mark above sinus, second mark above diffusing towards base. Ruslan Mishustin, Ukraine. 2021. Autumn/Winter.

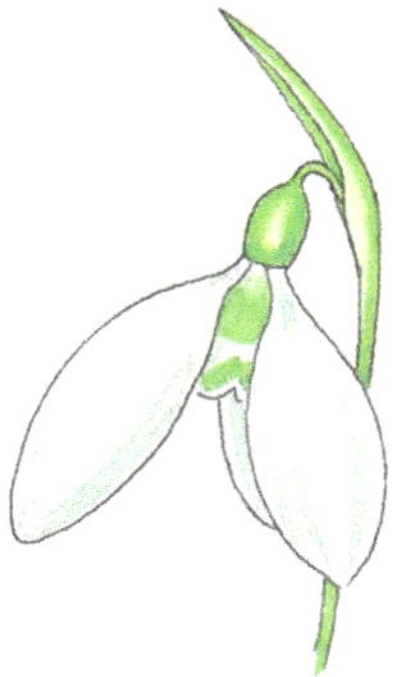

'Alpha Children of Triffids'
Very slender, green-marked flowers with split spathe and long slender green ovary. Outer segments slender, incurved, tapering to pointed apex, green mark above apex. Inner segments, green inverted 'V'-shaped mark above sinus, second elongated mark towards base. Ruslan Mishustin, Ukraine.

'Alpha Chinese Lantern'
Very rounded, green-marked Inverse Poculiform flowers. Leaves broad, erect, lightly longitudinally ridged, shiny green. Outer segments broad, rounded, longitudinally ridged, rounded to apex, broad green mark above small sinus. Inner segments green mark above sinus. Ruslan Mishustin, Ukraine. 2021.

'Alpha Christmas Bells'
Smaller, early snowdrop. Outer segments spoon shaped, tapering to bluntly pointed apex. Inner segments squared green mark above sinus, second mark above diffusing towards base. Flowers appear before leaves. October-November. Ruslan Mishustin, Ukraine. 2021. 10cm.

'Alpha Compliment'
Heavily green marked, boat shaped flowers on erect scapes with dark green spathe and triangular ovary. Outer segments rounded to bluntly pointed apex, longitudinally ridged with green stripes running up ridges from above apex to base, merging into green mark at base. Ruslan Mishustin, Ukraine.

'Alpha Dachshund'
Autumn-flowering *G. bursanus* with slender flowers. Outer segments slender, tapering to pointed apex. Inner segments two green marks, small inverted 'V' with broad ends above sinus, two joined ovals towards base. Flowers appear before leaves. 2021. Ruslan Mishustin, Ukraine. October-November. 15cm.

'Alpha Detective Poirot'
Erect stems with slender spathe and small rounded dark-green ovary. Outer segments long, slender, rounded, lightly longitudinally ridged, Inner segments narrow inverted green 'V'-shaped mark above sinus, second mark towards base. Ruslan Mishustin,Ukraine.

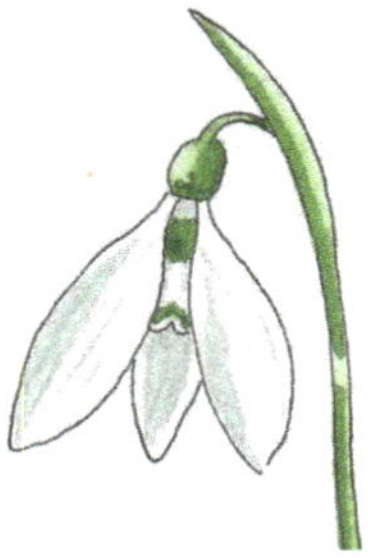

'Alpha Droopy'
Rounded, heavily green-marked flowers suspended from slender pedicel and oblong green ovary. Outer segments broad, rounded to bluntly pointed apex, merging green lines above apex to half of segment. Inner segments green 'W'-shaped mark above sinus and two small green ovals towards base. Ruslan Mishustin, Ukraine.

'Alpha Elhaz'
Early slender flowers with slender rounded ovary. Outer segments slender, rounded, tapering to pointed apex. Inner segments double mark, inverted 'V' above sinus with squared ends. Oval mark above diffusing towards base. Leaves short at flowering. Ruslan Mishustin, Ukraine. 2021. October-November. 15cm.

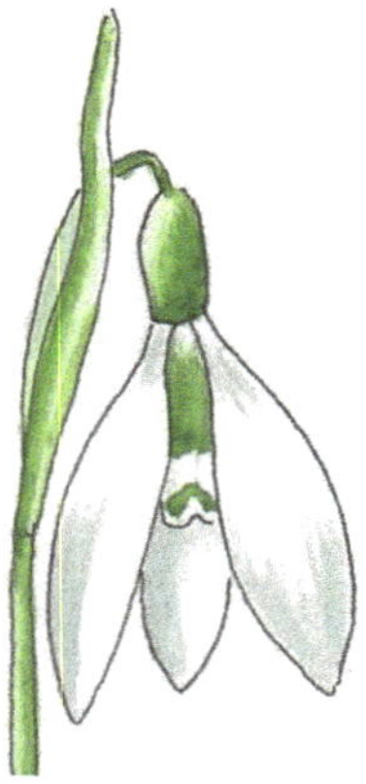

'Alpha English Lawn'
Rounded flowers beneath small rounded green ovary on erect scapes with slightly inflated spathe. Outer segments rounded, rounded at apex, longitudinally ridged, small dark green lines above apex. Ruslan Mishustin, Ukraine.

'Alpha First Snow'
G cilicicus. Rounded flowers with good heart-shaped mark and small rounded dark-green ovary. Leaves applanate, glaucous. Outer segments rounded, rounded at apex and lightly textured. Inner segments good green heart-shaped mark above indistinct sinus. November-December. Ruslan Mishustin, Ukraine.

'Alpha Flying Birds'
Full, well-rounded flowers. Outer segments rounded tapering to pointed apex, green lines above apex. Inner segments broad filled in inverted 'U'-shaped mark above sinus to half of segment. Two smaller, paler 'eye' dots towards base. Ruslan Mishustin, Ukraine.

'Alpha Gold n Pearls'
Shorter-growing snowdrop with very slender, pointed flowers, yellow-tinged spathe, yellow pedicel and ovary. Outer segments slender tapering to very pointed apex. Inner segments yellow mark above sinus diffusing on basal edge. Ruslan Mishustin, Ukraine. Winter/ Spring. 8-10cm.

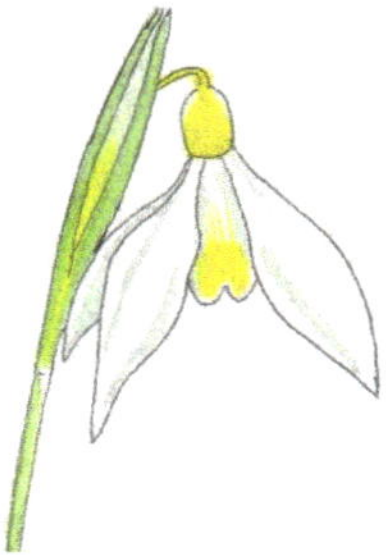

'Alpha Guppies'
Shining, almost translucent flowers on erect scapes with slender segments. Outer segments slender, tapering to pointed apex, light longitudinal ridging. Inner segments, narrow with slender green heart-shaped mark above sinus. Ruslan Mishustin, Ukraine. January-February.

'Alpha Helmet / Alpha Teutonic Helmet'
Rounded flowers beneath rounded green ovary. Outer segments rounded, bluntly pointed at apex, lightly longitudinally ridged, slightly flared. Inner segments green mark either side of sinus, second green mark towards base. Ruslan Mishustin, Ukraine.

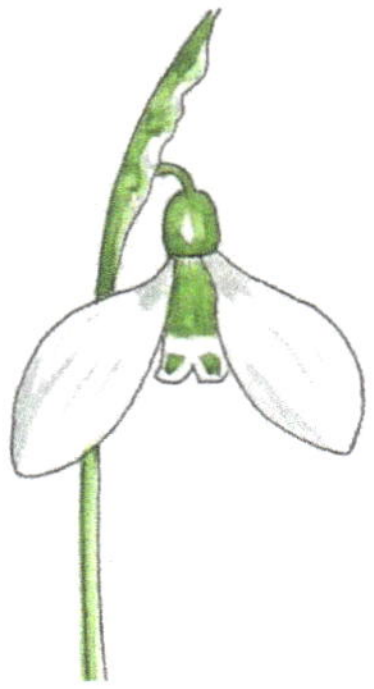

'Alpha Hope'
Smaller snowdrop with a cluster of slender green-marked segments in outward facing flowers. Segments five-six, narrow with green mark above apex. Ruslan Mishustin. Ukraine. 2019. 10cm.

'Alpha Humpty Dumpty'
Early flowering *G. cilicicus* with very broad rounded flowers beneath small rounded dark-green ovary. Leaves applanate, glaucous. Outer segments well rounded, rounded at apex and lightly textured. Inner segments, broad inverted filled in 'U'-shaped mid-green mark above sinus.

'Alpha Husky'
Autumn-winter flowering *G. bursanus.* Leaves blue-green, small at flowering. Outer segments lightly longitudinally ridged. Inner segments double marks, small triangular mid-green mark either side of sinus, two mid-green ovals towards base. Ruslan Mishustin, Ukraine. 2021. October-November. 14cm.

'Alpha Icicles'
Leaves erect to splayed, slender, blue-green. Outer segments boat-shaped rounded at apex. Inner segments lemon-green inverted 'U'-shaped mark above sinus bleeding slightly on basal side. Ruslan Mishustin, Ukraine. Winter/Spring. 2021. 12cm.

'Alpha Jade Comb'
Heavily green-marked flowers on erect scapes. Outer segments short claw, rounded to bluntly pointed apex, green lines above apex to three quarters of segment. Inner segments green mark above sinus to base, narrow white margin. Ruslan Mishustin, Ukraine. 2019.

'Alpha Kiwi Mousse'
Good green-marked flowers with rounded ovary and erect scapes. Leaves broad erect, grey-green. Outer segments tapering to bluntly pointed apex, washed green. Inner segments green mark. Ruslan Mishustin, Ukraine. Winter/Spring. 10cm.

'Alpha Ladybug'
Attractive *G. bursanus* with well rounded flowers on erect scapes beneath slender light-green ovary. Outer segments rounded, tapering to pointed apex, longitudinally ridged. Inner segments narrow pale-green inverted 'V'-shaped mark with rounded ends above sinus, second pale-green mark mid-way, extending almost to base. Ruslan Mishustin, Ukraine. 2021. Autumn-winter flowering.

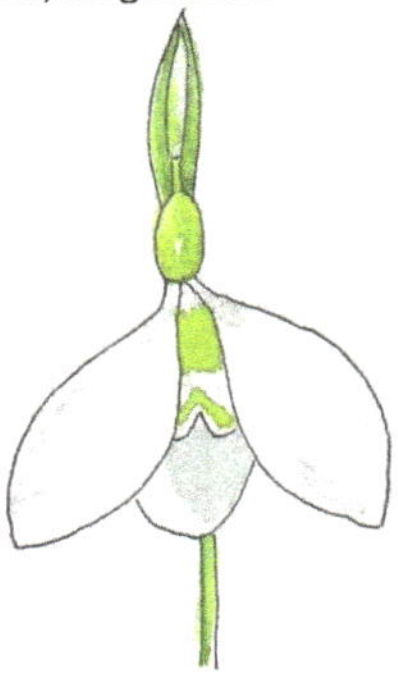

'Alpha Light Nostalgia'
Long, slender, green-marked flowers. Outer segments long, slender tapering to pointed apex, green lines above apex on lower third of segment. Inner segments broad green mark above sinus. Ruslan Mishustin, Ukraine.

'Alpha Little Mermaid'
Rounded, green-marked flowers with oblong green ovary. Outer segments rounded, tapering to pointed apex, green lines above apex to half of segment, green wash. Inner segments green mark above sinus towards base, narrow white margin. Ruslan Mishustin, Ukraine.

'Alpha Merlin's Brushes'
G. bursanus. Smaller snowdrop with rounded dark-green ovary, Leaves small at flowering, erect, blue-green. Outer segments rounded, tapering to bluntly pointed apex, green lines above apex to half of segment. Inner segments good green mark from above sinus to base. November-December, Ruslan Mishustin, Ukraine.

'Alpha New Art'
Elegant-looking Inverse Poculiform flowers with good green marks. Leaves erect, slender, blue-green. Outer segments long, incurved, rounded to bluntly pointed apex, light green mark above apex to one third of segment. Inner segments shorter than outer, light-green heart-shaped mark above sinus. Ruslan Mishustin, Ukraine. 2021. Winter-Spring. 16cm.

'Alpha Old Joke'
Slender flowers. Outer segments slender, rounded, bluntly pointed at apex. Inner segments narrow green inverted 'V'-shaped mark above sinus joining slender, paler stripe to base. Ruslan Mishustin, Ukraine.

'Alpha Paradox'
Long, slender flowers on erect scapes. Leaves erect, slender, blue-green. Outer segments long, slender bluntly pointed at apex. Inner segments double mark, inverted 'V'- shape above apex, rounded mark towards base. Ruslan Mushustin, Ukraine. 2022.

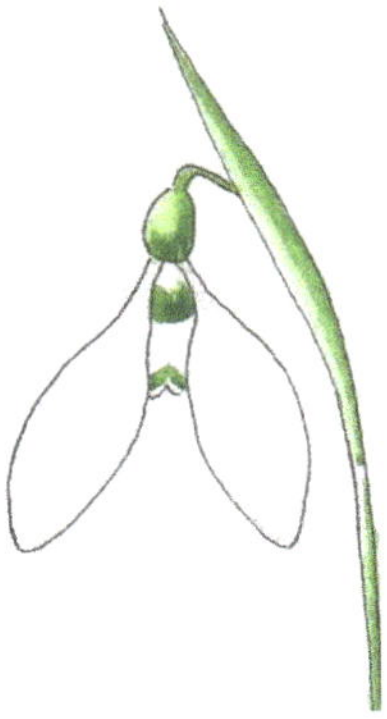

'Alpha Pistachio Ice Cream'
Rounded flowers beneath small green ovary. Outer segments rounded, lightly longitudinally ridged, rounded at apex, mid-green stripes above apex to half of segment. Inner segments green mark above sinus to base, diffusing towards base. Ruslan Mishustin, Ukraine.

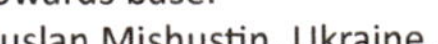

'Alpha Point'
Erect scapes with rounded flowers beneath small, rounded green ovary. Outer segments rounded, lightly longitudinally ridged, tapering to pointed apex, small, pale-green dot on lower third of segment. Inner segments filled in, inverted 'U'-shaped mark above sinus. Ruslan Mishustin, Ukraine.

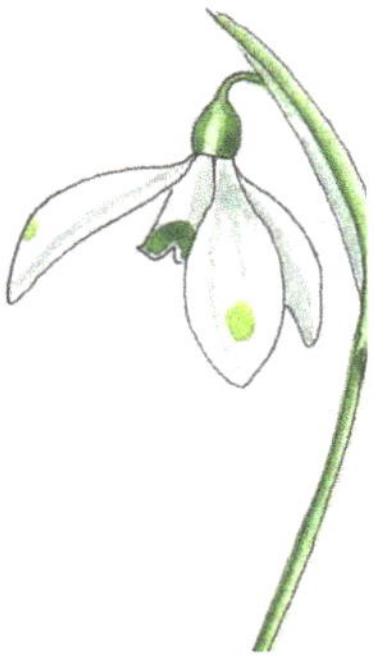

'Autumn Praying Hands'
Autumn-flowering *G. elwesii monostictus* with upward facing, rounded flowers tapering to rounded apex, and erect Scharlocki type split spathe. Inner segments, narrow mid-green inverted 'V'-shaped mark above sinus, second mid-green oval towards base. Ruslan Mishustin, Ukraine.

'Alpha Predator'
G. plicatus with rounded, outward facing flowers and oblong mid-green ovary. Leaves explicative, green-grey-green. Outer segments broad, rounded, tapering to outward turning, bluntly pointed apex, strong mid-green lines above apex towards base. Inner segment green mark. Ruslan Mishustin, Ukraine.

'Alpha Pug Dog'
Quirky looking, smaller, rounded flowers with green marked segments and large oblong, rounded green ovary. Outer segments rounded, variably marked green at apex and base. Inner segments green mark. Ruslan Mishustin, Ukraine. 8-10cm.

'Alpha Roly Poly'
Great-looking *G. bursanus* snowdrop with weighty, rounded flowers beneath small rounded green ovary. Outer segments broad, rounded, rounded at apex. Inner segments small green mark either side of sinus, two narrow paler green marks towards base. Ruslan Mishustin, Ukraine. Autumn-Winter.

'Alpha Seven Dwarfs'
Large flowers on short erect scapes with oblong green ovary. Leaves erect, green. Outer segments broad, tapering to pointed apex, longitudinally ridged with small green lines above apex. Inner segments broad green mark above sinus to half of segment. Ruslan Mishustin, Ukraine. Winter/Spring. 8cm.

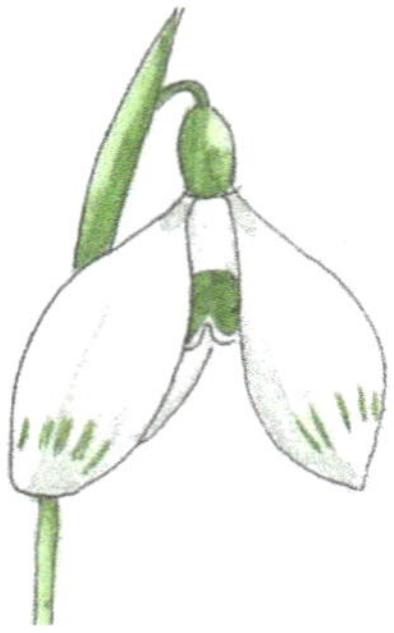

'Alpha Spring Crinoline'
Bold, well-rounded, heavily textured, balloon-like flowers suspended from long slender pedicel. Outer segments very broad, rounded to rounded apex, strongly longitudinally ridged and dimpled, green lines on lower half of segment running up ridges. Inner segments, green mark above sinus to base, narrow white margin. Ruslan Mishustin, Ukraine. 2016. Winter/Spring. 18cm.

'Alpha Striped Beauty'
Attractive, rounded green-marked flowers. Outer segments broad, rounded to bluntly pointed apex, light-green lines above apex to three quarters of segment. Inner segments light-green mark above sinus almost to base, narrow white margin. Ruslan Mishustin, Ukraine.

'Alpha Striped Skittles'
Rounded flowers beneath slender green ovary. Leaves slender, grey-green, short at flowering. Outer segments rounded, tapering to bluntly pointed apex. Green lines above apex. Inner segments solid green mark from above sinus almost to base, narrow white margin. Ruslan Mishustin, Ukraine.

'Alpha Suns Favourite'
Smaller snowdrop with greenish-yellow scape, spathe and pedicel, rounded flowers and very distinctive, variegated foliage. Leaves medium, spreading, variegated, yellow-green. Outer segments rounded to bluntly pointed apex. Ruslan Mishustin, Ukraine. Winter/Spring. 8-10cm.

'Alpha Tutelar'
Quirky, small autumn-flowering *G. bursanus*, with short scapes and long yellow-green ovary. Leaves slender erect, short at flowering. Outer and inner segments of equal length, flaring, rounded, rounded at apex, lightly longitudinally ridged. Inner segments narrow inverted 'U'-shaped mark above small sinus. October. Ruslan Mishustin, Ukraine.

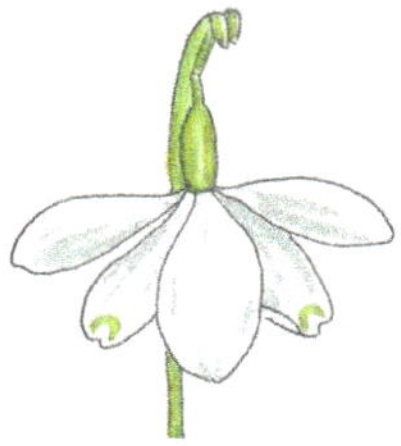

'Alpha White Claws'
Slender flowers with narrow green ovary. Outer segments slender, rounded, tapering to pointed apex. Inner segments double green mark, narrow inverted 'U'-shaped green mark above sinus, broad oval green mark towards base. Ruslan Mishustin, Ukraine. 8cm.

'Alpha White Dragon'
Large, pure-white flowers on erect scapes held well above the leaves. Leaves slender, erect, blue-green, short at flowering. Outer segments broad, rounded to bluntly pointed apex, lightly longitudinally ridged. Inner segments broad, pure white, small sinus. Ruslan Mishustin, Ukraine. Autumn/Winter. 15cm.

'Alpha Yellow Fairy'

Rounded flowers with yellow-green pedicel and ovary. Outer segments rounded, tapering to bluntly pointed apex, lightly textured. Inner segments yellow-green inverted 'V'-shaped mark above sinus, diffusing on basal edge. Ruslan Mishustin, Ukraine. Winter/Spring.

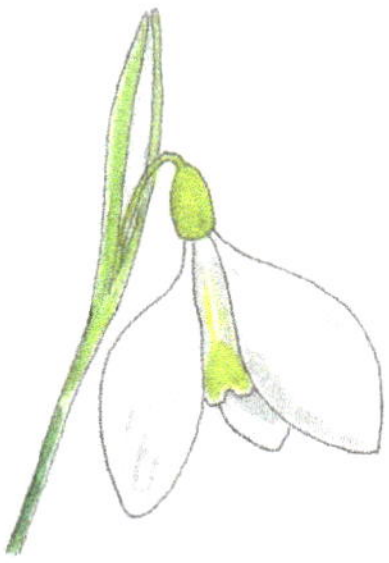

'Alpha Yeti'

Outer segments slender, rounded to pointed apex, lightly longitudinally ridged. Inner segments narrow green inverted 'U'-shaped mark above sinus, second oval towards base diffusing on basal edge. Ruslan Mishustin, Ukraine.

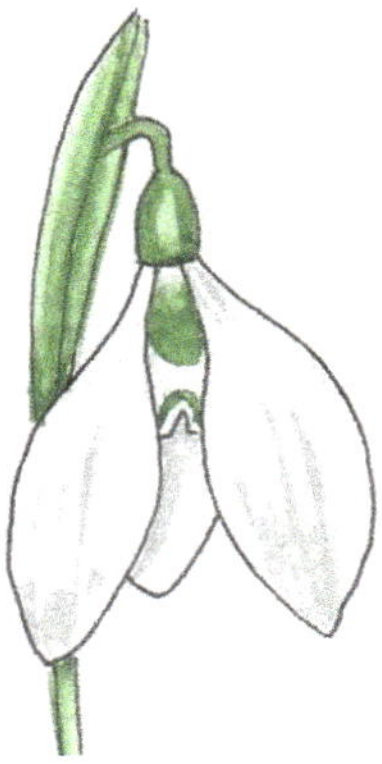

'*alpinus* v. *alpinus*'

Syn. G. caucasicus

Small plants from the mountains of Turkey and Russia. Leaves, slender, glaucous, usually short at flowering. Outer segments rounded, tapering to pointed apex. Inner segments small green inverted 'V'- or 'U'- shaped mark above sinus. Can be difficult and slow. Winter/Spring. 9-10cm.

'Altered Image'

G. elwesii var. *monostictus* with slender, rounded, variable flowers and rounded green ovary. Leaves supervolute, erect to arching, broad, grey-green-glaucous. Outer segments vary between six and three, incurved, tapering to pointed apex. Inner segments neat with green mark above apex. This snowdrop originated from a bulb with four by four flowers. Later and in mature bulbs, three by three, four by four and five by five flowers can be produced. Jöerg Lebsa, from his garden in Dresden. 15cm.

"Amy Doncaster's Double'

A short, rarely obtainable cultivar with full double flowers and broad, green-marked inner segments. Outer segments broad, incurved to bluntly pointed apex. Inner segments broad ruff with green marks above sinus. From the garden of the late Margaret Owen, Acton Pigot, Shropshire. 10cm.

'Angel Fley'

Small, neat and pretty later-flowering snowdrop. Leaves short, green. Outer segments rounded to bluntly pointed apex, neat light-green lines above apex, light green shading at base. Inner segments inverted mid-green 'V'-shaped mark above sinus. Spring. 8cm.

'Angel Lady'

Good, strongly green-marked *G. nivalis*. Leaves applanate, erect, slender, blue-green. Outer segments rounded to bluntly pointed apex, merging green lines above apex to base. Inner segments, green mark above sinus to base, narrow white margin. 2020. Winter/Spring. 10cm.

'Angel of the North'

G. nivalis with slender yellow ovary, long pedicel and slender-looking, yellow-marked flowers. Leaves applanate, slender, erect, grey-green. Outer segments slender, tapering to pointed apex. Inner segments yellow inverted 'V'-shaped mark above sinus. Freddy Van Houtte. 2018. Winter/Spring. 10cm.

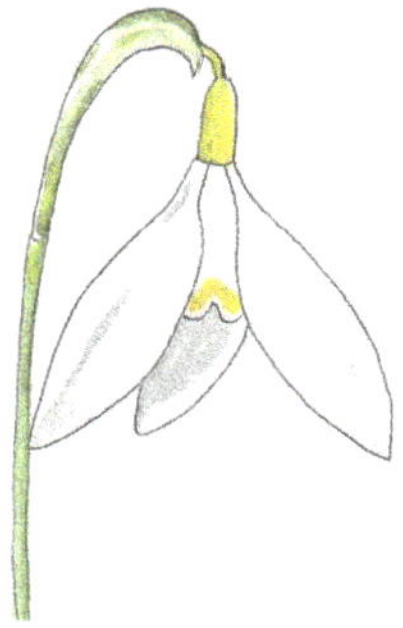

'Angel or Demon'

Quirky *G. nivalis* with odd looking flowers, twisted segments and small, triangular green ovary. Leaves applanate, slender, erect to arching, blue-green, paler median line. Outer segments variously twisted, pinched at apex, lightly longitudinally ridged. Inner segments variable, small green inverted 'V'-shaped mark above sinus.

'Angela's Gift'

Clean white-looking rounded flowers with oblong rounded green ovary. Outer segments broad, rounded, tapering to pointed apex. Inner segments narrow 'W'-shaped mark above sinus and a 'V'-shaped mark mid way up segment.

'Anglesey Cloudgazer'

Upward facing double snowdrop, especially in the immature phase, with small, triangular green ovary. Outer segments slender, flaring upwards, small green marks above apex. Inner segments neat, regular ruff, tipped green at sinus, Named for a wooden structure at Anglesey Abbey designed by Richard Todd, enabling visitors to lie back and look at the sky. Rainbow Farm selection from Anglesey Abbey. 2021.

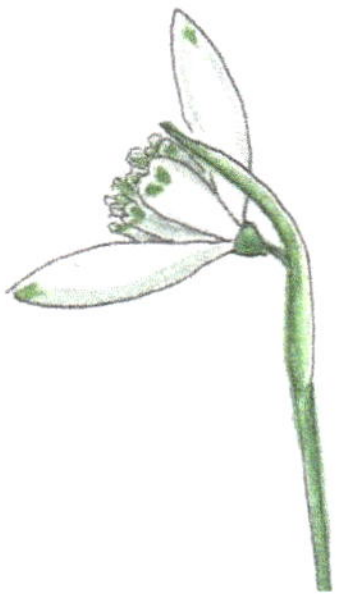

'Anglesey Doublet'

Rounded double flowers on slender pedicel with slender triangular green ovary. Outer segments rounded to bluntly pointed apex. Inner segments neat ruff, heavily green marked above sinus almost to base. Rainbow Farm selection from Anglesey Abbey.

'Anglesey Promise'
Good, well-rounded, fully double flowers on slender pedicel with small triangular green ovary. Outer segments broad, rounded to rounded apex. Inner segment ruff heavily marked green above sinus towards base, narrow white margin. Rainbow Farm selection from Anglesey Abbey.

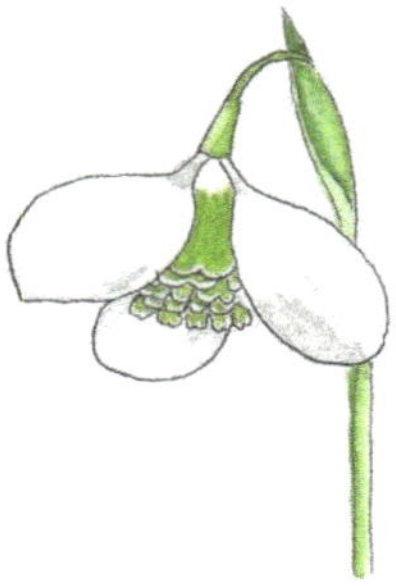

'Ann Baring'
Early-flowering hybrid with well-shaped flowers and dark-green ovary. Outer segments clawed, rounded, lightly longitudinally ridged and bluntly pointed at apex. Inner segments, strong slightly waisted dark-green inner mark from above sinus to base. It is said this mark resembles a diving fish. Spring.

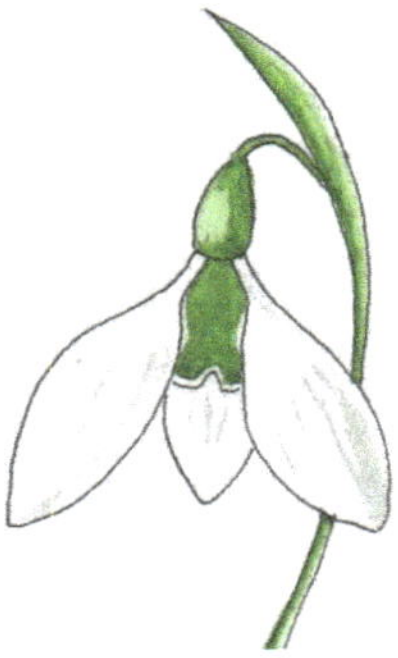

'Ann Christin'
Rounded flowers with oblong rounded green ovary. Outer segments broad, rounded to bluntly pointed apex. Inner segments, small green inverted 'V'-shaped mark above sinus, second green mark towards base.

'Annalivia'
Delicate looking flowers on erect scapes with small, yellow-green ovary. Leaves erect, broad, green, reflexed edges, central median ridge. Outer segments slender, rounded to bluntly pointed apex. Inner segments yellow-green heart-shaped mark above sinus, paler shadow towards base. Paddy Tobin in garden of Lyrath Hotel, Kilkenny and named for his granddaughters Olivia and Anna. Winter/Spring. 12cm.

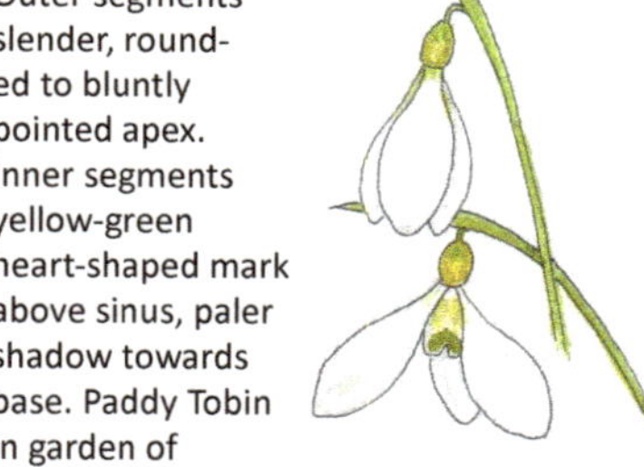

'Anna O'
G. nivalis with good yellow colouring on erect, yellow-green scape and spathe with long, slender, arching yellow pedicel and slender yellow-green ovary. Leaves applanate, narrow, spreading, pale blue-green. Outer segments rounded to bluntly pointed apex, lightly longitudinally ridged. Inner segments faintly flushed yellow with tiny yellow dot either side of sinus. Winter/Spring. 12cm.

'Anne's Pyjamas'
Slender, erect *G. elwesii* with narrow, pointed flowers. Leaves supervolute, erect, broad, grey-green, Outer segments, slender, pointed at apex, merging green lines from above apex to half of segment.

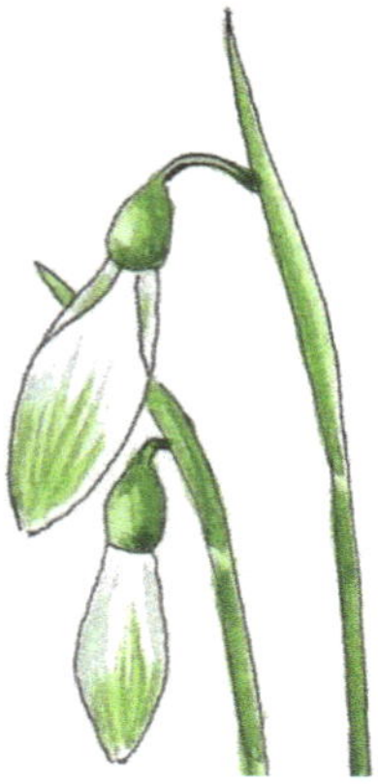

'Antares'
Good *G. nivalis* with green-marked flowers and oblong green ovary. Leaves applanate, splayed, slender, blue-green. Outer segments long, slender, rounded to bluntly pointed apex, bold mid-green mark above apex almost to base. Inner segments solid mid-green mark across segment from above sinus, almost to base, narrow white margin. Winter/Spring. 2020. 14cm.

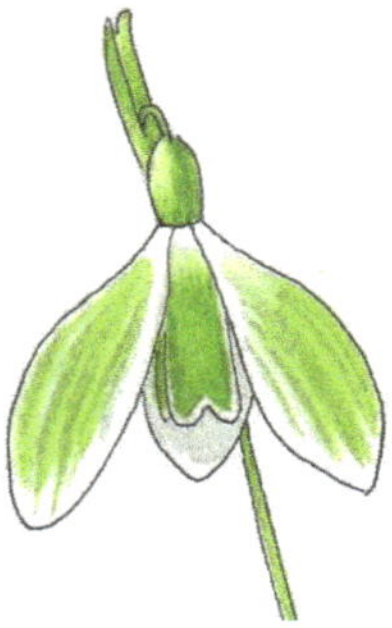

'Apatite'
A good, strong virescent snowdrop from a cross between *G.* 'Nova Gorica' and *G.* 'Green Tear'. Leaves slender, erect to arching. Outer segments long, rounded to apex, green lines above apex diffusing towards base, light-green wash.

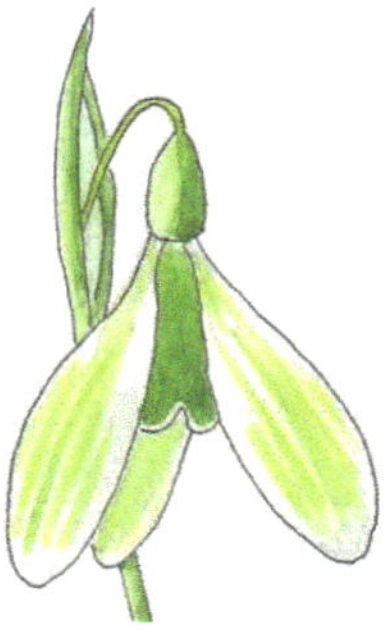

Inner segments mid-green mark from above sinus to base, narrow white margin. A good strong grower. Edulis Nursery. 2022. 16cm.

'Apollo'
Large rounded flowers on short, erect scapes. Leaves slender, erect, blue-green. Outer segments broad, rounded to pointed apex, lightly longitudinally ridged. Inner segments narrow green inverted 'V'-shaped mark above sinus diffusing into yellow on basal side. Winter/Spring. 10cm.

'Appleby Spiky'
G. nivalis with spiky double flowers in an untidy cluster of long, slender, lightly green-marked segments and small, slender, triangular green ovary. Raised by the late Hector Harrison, Appleby, Lincolnshire and originally stocked by Pottertons Nursery, c 1990s.

'Arabian Gold'
An elegant yellow *G. nivalis* on erect yellow-green scape with yellow pedicel and ovary. Leaves applanate, erect, slender, green, paler median line. Outer segments slender, tapering to pointed apex. Inner segments, yellow inverted 'V'-shaped mark above sinus. Good grower and bulks up well. Freddy Van Houtte. Winter/Spring. 8cm.

'Ariadne'
G. plicatus subs. *byzantinus*. Slender pedicel with oblong green ovary. Leaves explicative, green-grey. Outer segments clawed, rounded to bluntly pointed apex. Inner segments inverted green 'V'-shaped mark above sinus and second green mark towards base. Will eventually produce two flowers from a single scape. Named for the Cretan Goddess married to Theseus and associated with labyrinths, and the bearer of twins. Winter/Spring. Avon Bulbs.

'Arto's Secret'
A cross between *G.* 'Trym' and *G.* 'Sutton Courtney' from Joe Sharman giving a tall, neat inverse Poculiform snowdrop. Outer segments flattened, rounded towards rounded apex. Broad green heart-shaped mark above apex. Inner segments green mark above sinus diffusing on basal side. Joe Sharman, Monksilver Nursery, Cambridge and named for the grandson of a friend. 2022. Winter/Spring. 20cm.

'Ate Tea'
Beautifully shaped *G. rizehensis* with elegant flowers on erect scapes with long green ovary. Leaves applanate, linear, slender, erect, blue-green, lighter median line. Outer segments slender, rounded to bluntly pointed apex. Inner segments slender with good, bright green mark above sinus almost to base. Andy Byfield. Winter/Spring. 8cm.

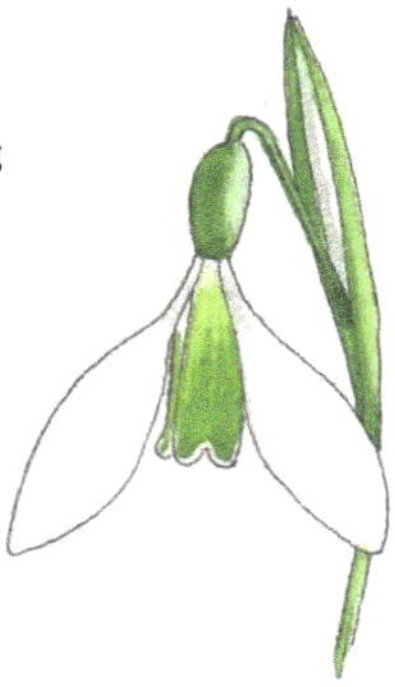

'Atkinsii Variegated'
Leaves erect to splayed, broad, green, with attractive variegated cream stripes, some plants have stronger variegation than others. Outer segments rounded to bluntly pointed apex. Inner segments green inverted 'U'-shaped mark above sinus. Vigorous. 2009. Winter/Spring. 16cm.

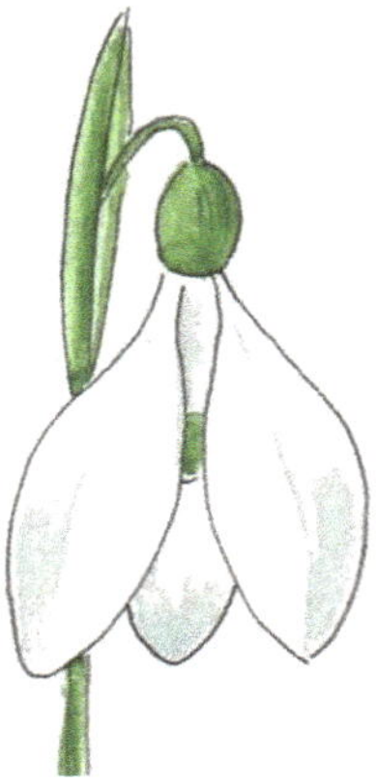

'Augmented Triad'
Erect scapes and rounded ovary. Outer segments rounded, tapering to pointed apex. Inner segments, broad green inverted 'V'-shaped mark above sinus. Johan Mens. 2022. An augmented triad is a musical term.

'August Der Starke'
('August The Strong') Balloon-like, well rounded, textured flowers on erect scapes with small round green ovary. Outer segments, broad, rounded, rounded at apex, lightly longitudinally ridged, textured. Inner segments green mark above sinus diffusing towards base. German origin.

'Aul'
Recent introduction in the attractive Estonian Bird series of double snowdrops. ('Aul' = Long Tailed Duck - *Clangula hyemalis*) Erect scapes with long, slender, arching pedicel merging into slender triangular ovary. Outer segments rounded, slender claw, bluntly pointed at apex. Interesting arrangement of Inner segments, broad with slender inverted green 'V'-shaped mark above sinus, pale green wash above. Taavi Tuulik, Estonia. 2020.

'Aunt Nellie Danglers'
Long, large, slender flowers on short scapes. This clone of *G. elwesii* originated in a garden in Berkshire on the Midvale Ridge. Outer segments very long, up to 50mm, slender, tapering to pointed apex and three times longer than inners. Inner segments double mark, good heart-shaped green mark above sinus and second mark towards base. The name comes from a slang term for earrings. A number of other snowdrops have been found in the same area of the Midvale Ridge. Andy Byfield, Flete Walled Garden, South Devon.

'Aztec Gold'
Good strong gold colouring with pedicel, and ovary on this striking *G. nivalis*. Leaves applanate, medium. splayed, grey-green. Outer segments broad, clawed, rounded to bluntly pointed apex, lightly longitudinally ridged. Inner segments, strong gold inverted 'U'-shaped mark above sinus. Ruben Billiet. 10-15cm. Winter/Spring.

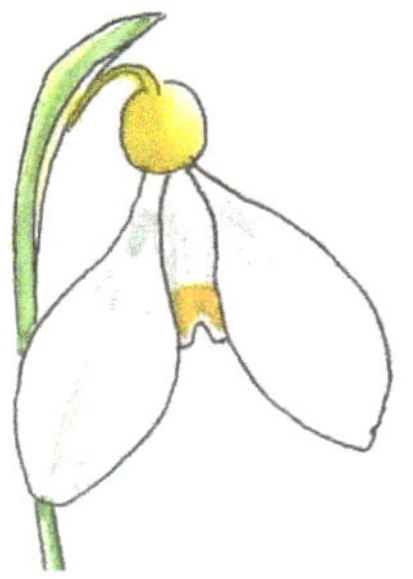

'Babraham Braveheart'
Outer segments rounded, tapering to bluntly pointed apex. Inner segments green inverted 'V'-shaped mark above sinus with rounded ends. Rainbow Farm. 2021.

'Bad Hair Day'
Small, quirky *G. nivalis* with full, flaring, rounded flowers and small triangular green ovary. Leaves, applanate, short, slender, blue-green. Outer segments long, slender, bluntly pointed at apex with small green mark at apex on some segments. Inner segments Inverted green 'V'-shaped mark above sinus. Ruben Billiet. Belgium. Winter/Spring. 8cm.

'Badger's End'
G. rizehensis with flaring flowers and small rounded green ovary. Leaves applanate, slender, erect, mid-green. Outer segments flaring upwards, slender, waved, bluntly pointed at apex. Inner segments, pale-green inverted 'V' shaped mark above sinus with rounded ends. 2019. Winter/Spring. 10cm.

'Ball'
Well rounded double flowers with variable green marks. Outer segments rounded, incurved. Inner ruff tightly packed with variable green marks above sinus.

'Bambino'
As the name implies, small plants with large flowers beneath curving spathe, yellow/green ovary and pedicel. Leaves long, spreading. Outer segments broad, rounded to pointed apex. Inner segments narrow green inverted 'V'-shaped mark above sinus. 8cm.

'Barleymow'
Attractive green-tipped *G. nivalis* with oblong, rounded green ovary on erect scapes. Leaves applanate, erect to arching, slender, blue-green. Outer segments tapering to pointed apex, small green mark above apex. Inner segments green mark above sinus. Edulis. 2020. Winter/Spring. 15cm.

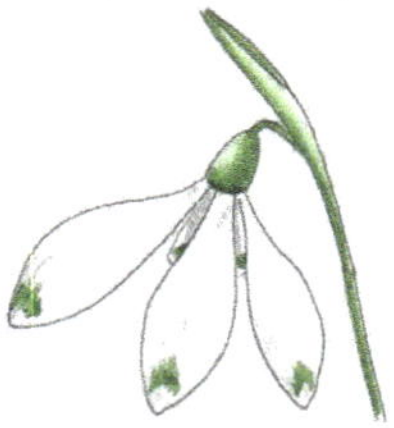

'Basket'
A rather strange looking *G nivalis* with flowers held at right angles to the scape and spathe, as well as often throwing segments from the ovary. Decora Nursery, Croatia. 2020. Winter/Spring. 10cm. (No illustration)

'Be My Valentine'
Inverse Poculiform from *G.* 'Trym' *x G.*'South Hayes' with long flowers beneath rounded ovary. Outer segments flattened, rounded to apex, broad green mark from above good sinus almost to base. Inner segments good green mark above sinus diffusing towards base. Valentin Wijnen. 2019. Winter/Spring, usually in time for Valentine's day.

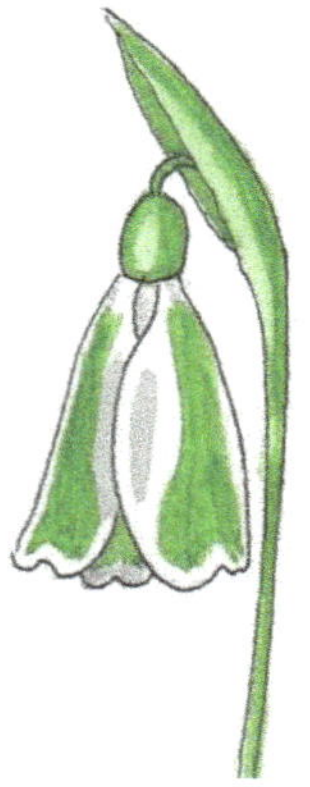

'Bead Bear'
Beautiful rounded looking Poculiform flowers with all white segments and rounded green ovary. Leaves erect, broad, grey-green. Segments all of equal length, rounded to apex, white. Cal Mateer, British Columbia.

'Beekeeping Cream of Bauerlein'
G. nivalis with erect scapes, narrow spathe and open, spreading flowers with angled pedicel and small oblong, rounded ovary. Leaves applanate, slender, erect to arching, blue-green. Segments all roughly of equal length, slender, tapering to bluntly pointed apex with cream shadow at base of segments. Winter/Spring. 12cm.

'Belgium Trust'
A semi-double *G. nivalis* with small green triangular ovary. Leaves applanate, erect, slender, blue-green. Outer segments flaring, rounded, tapering to bluntly pointed apex. Inner segments variable 2x3 with the outer whorl having a broad inverted green 'V'-shaped mark above sinus. The second whorl is slender, rounded at apex with variable small green mark above apex. Freddy Van Houtte. 14cm.

'Belisha Beacon'
Attractive, heavily green marked stately looking snowdrop with oblong green ovary and broad, grey-green leaves. Outer segments, short claw, pronounced shoulder, broad, rounded, tapering to pointed apex. Inner segments narrow, flared at apex, narrow green inverted 'V'- shaped mark above sinus, second green mark from half way up segment to base. Winter/Spring.

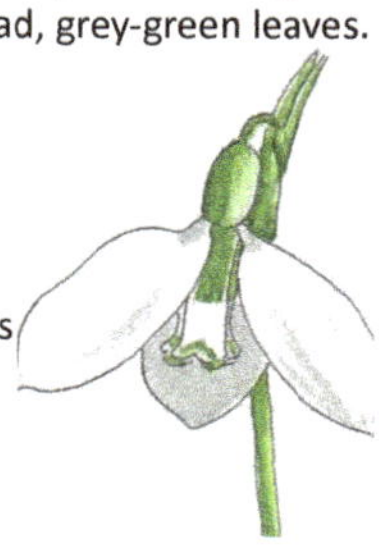

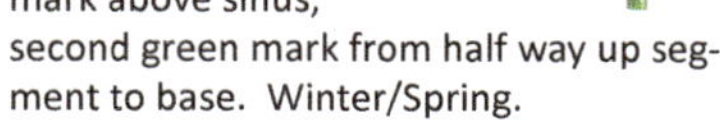

'Bell Star'
Small-flowered *G. nivalis* Inverse Poculiform snowdrop with segments roughly of equal length. Outer segments flattened, rounded to rounded apex, green mark above small sinus. Inner segments, mark either side of sinus. Found in an abandoned manor house garden in Normandy, France, by Belgian Galanthophile Oliver Vico and his wife Trudy.

'Bella Bianco'
Slender-looking *G. nivalis* on erect scape with long, slender pedicel and slender ovary. Leaves applanate, erect to arching, slender, blue-green. Outer segments long, slender, rounded to pointed apex. Inner segments thin inverted green 'V'-shaped mark above sinus. Gerard Oud. 15cm.

'Bend Bear'
Interesting *G. elwesii* Poculiform flowers. All segments roughly of equal length, rounded, strongly reflexing upwards from short claw, lightly textured and longitudinally ridged, rounded at apex. light-green splash near apex. Cal Mateer. British Columbia. 2022.

'Beta Galanthopath'
Weighty-looking, well-rounded flowers with small triangular ovary on a shorter growing snowdrop. Leaves broad, lightly ridged, grey-green. Outer segments short claw rounded and rounded at apex. Inner segments good green mark to base. 10cm.

'Betelgeuze'
Long, slender Inverse Poculiform flowers. Leaves erect, broad, green with central rib. Outer segments long, slender tapering to pointed apex, broad green mark above apex to half of segment and often in a thin line to base. Inner segments broad green mark above sinus to half of segment.

'Bethany'
Good *G. nivalis x G. plicatus*, virescent with slender dark-green ovary. Leaves medium, erect to arching, good mid-green. Outer segments taper to pointed apex, green lines above apex paling at base. Inner segments green mark across segment from above sinus almost to base, narrow white margin. Paul Barney, Edulis Nursery 2021.

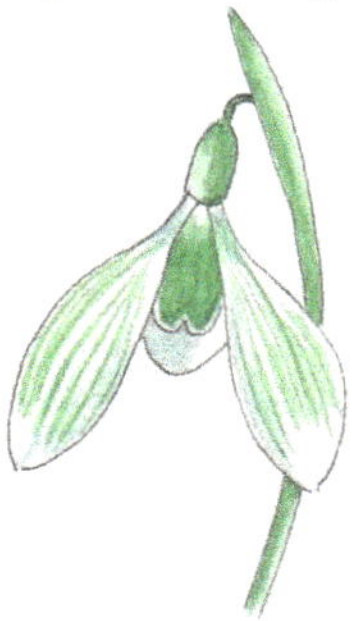

'Betty Hughes'
Well-rounded, very neat, double Hybrid on slender arching pedicel, with good green inner mark. Leaves medium, erect to splayed, green, paler median line. Outer segments broad, rounded, rounded to apex. Inner segments neat ruff with one segment protruding like a proboscis from the centre, broad inverted green 'V'-shaped mark above sinus lightly joining two small merging ovals towards base Joe Sharman, Monksilver Nursery. 2020. Winter/Spring. 15-16cm.

'Big Ben'
Elegantly shaped flowers with slender yellow-green pedicel and ovary. Outer segments long, clawed, rounded and tapering to pointed apex. Inner segments slender, yellow inverted 'U'-shaped mark above sinus.

'Big Egg'
Large rounded flowers with rounded green ovary. Outer segments broad, rounded to bluntly pointed apex, lightly longitudinally ridged.
Inner segments green inverted 'V'-shaped mark above sinus, the mark dividing and paling upwards to base. above sinus.

'Big One'
Large weighty looking flowers on this attractive short *G. elwesii* clone. Leaves broad, erect, explicative, grey-green. Outer segments rounded to bluntly pointed apex. Inner segments narrow green inverted 'U'- shaped mark above sinus. Matt Bishop.

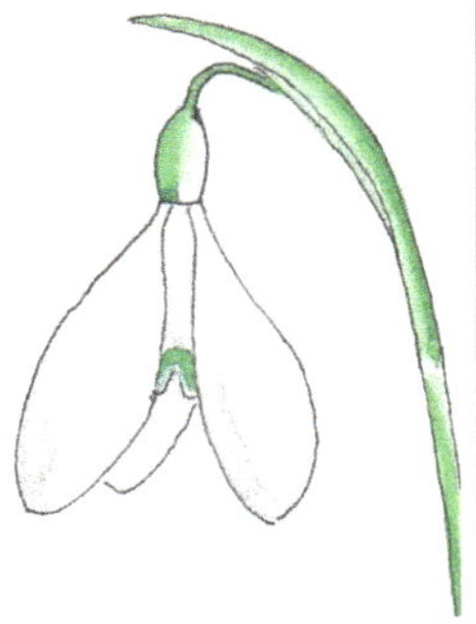

'Bitter Lemons'
a *G. x valentinei* hybrid with attractive yellow marking, starting yellow-green and deepening to gold as flowers mature. Leaves medium, erect to arching, blue-green. Outer segments short claw, rounded to bluntly pointed apex, yellow mark above apex. Inner segments inverted yellow 'V' to 'W'-shaped mark above sinus, deepening to gold. Found at the same site as G. 'Midas' but much earlier flowering. Horst Bauerlein.

'Bledlow Double'
G. nivalis found in Bledlow village, Buckinghamshire. (No illustration)

'Blond Birgit'
Delicate looking flowers suspended from long arching pedicel with rounded green ovary. Outer segments rounded to bluntly pointed apex. Inner segments, pale yellow inverted 'U'-shaped mark above sinus.

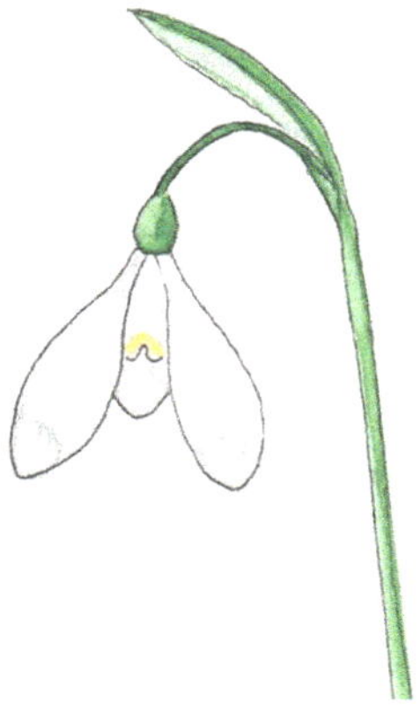

'Blonde Isabel'
Flaring outer segments reveal yellow marked inner segments. Outer segments rounded, incurved, tapering to bluntly pointed apex. Inner segments yellow mark above sinus.

'Blue Jay'
Attractive *G. plicatus* with well-rounded flowers with rounded ovary. Leaves explicative, medium, silver-grey, paler median line. Outer segments broad, rounded to bluntly pointed apex, lightly longitudinally ridged, goffered at base. Inner segments green mark above sinus almost to base, diffusing slightly on basal edge.

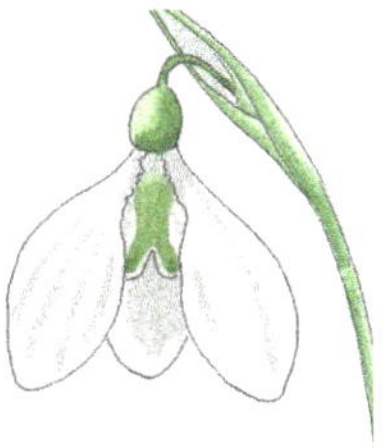

'Blue Octopus'
An early-flowering *G. elwesii* var. *monostictus*, giving substantial looking plants from large bulbs with large, well-shaped flowers, long green ovary and wide spreading foliage. Leaves supervolute, broad, long, lax, spreading - 'octopus like', distinctive blue-green and up to 45cm long in mature plants. Outer segments, rounded tapering to bluntly pointed apex. Inner segments single green mark above sinus. November-December. David Culp.

'Boat Bear'
G. elwesii Poculiform opening to rounded, white, boat-shaped segmented flowers, tapering to pointed apex. Leaves supervolute, broad, erect and reflexing, green-blue. Buds cream fading to white as flowers mature, tiny light-green marks at apex. Cal Mateer, British Columbia, 2021. Winter/Spring. 16cm.

'Bob's Seedling'
Rounded flowers on erect scapes with rounded green ovary. Outer segments rounded to bluntly pointed apex. Inner segments narrow green inverted 'V'-shaped mark above sinus.

'Bonanaza' - 'BGYL2'
Tall *G. nivalis* with delicate looking flowers and slender, rounded ovary. Leaves applanate, arching, slender, green. Outer segments slender, rounded to pointed apex. Inner segments, narrow inverted green 'V'-shaped mark above sinus. Slow to bulk up. 2019. Winter/Spring.18cm.

'Bong Bear'
Good all-white Poculiform flowers suspended from slender pedicel with rounded green ovary. Outer segments broad, rounded to pointed apex. Inner segments of equal length, incurved.
Cal Mateer, British Columbia.

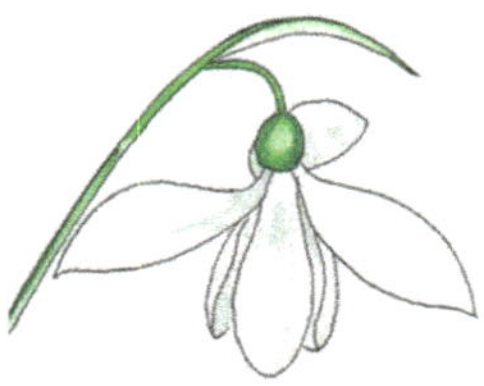

'Bowles Boys'
Hybrid which usually manages two flowers to each erect scape with long slender rounded yellow-green ovaries. Leaves erect, slender, blue-grey. Outer segments long claw, rounded to bluntly pointed apex, lightly longitudinally ridged. Inner segments green mark above sinus almost to base. Myddleton House.

'Bowles Gemini'
Mid-season Poculiform snowdrop often with two flowers from a scape. Leaves plicate, medium, green. Segments roughly of equal length, well rounded, tapering to bluntly pointed apex, lightly longitudinally ridged and puckered. This is a third Poculiform plicate from Myddleton House. Winter/Spring.

'Bowling Bear'
Attractive Poculiform flowers with segments of equal length. Similar looking snowdrop to G. 'E. A. Bowles'. Leaves erect to splayed, broad, grey-green. Segments rounded to pointed apex, white. Cal Mateer, British Columbia.

'Brechin Tower'
Rounded flowers with good inner mark and rounded green ovary. Outer segments incurved, rounded to pointed apex. Inner segments, slender, erect green mark above sinus to three quarters of segment. Ian Christie and named as the inner mark is said to resemble Brechin Cathedral. 2019. Winter/Spring. 15cm.

'Brian Ellis'
A good *G. nivalis* chosen when on an exploration of the Greatorex plot. Similar to *G.* 'Greenfinch', but more vigorous with a more substantial inner mark. Slender looking flowers suspended beneath long green ovary. Leaves applanate, erect to arching, grey-green. Outer segments long, slender, tapering to pointed apex, green lines above apex. Inner segments mid-green mark from above sinus almost to base, diffusing on basal edge. Found by Joe Sharman and Richard Hobbs, custodian of the plot, and named for Galanthophile Brian Ellis. 2020. Vigorous. Winter/Spring. 16cm.

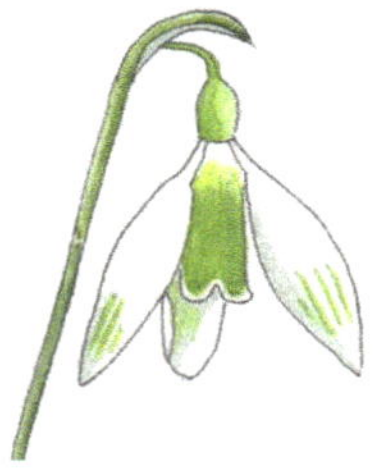

'Broad-Leaved Convolute'
Clean looking flowers suspended from slender pedicel with slender oblong green ovary. Leaves broad, convolute, grey-green. Outer segments rounded to bluntly pointed apex. Inner segments large green horseshoe shaped mark above sinus.

''Bullfight' ('Syn. Stiebritz')
Leaves broad, splayed, green. Outer segments light dimpling and goffering at base, tapering to bluntly pointed apex. Inner segments green mark above sinus almost to base, narrow white margin. Germany 2012.

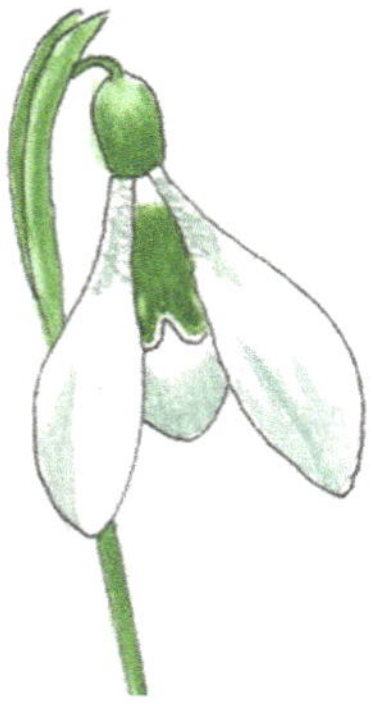

'Bunzaue'
A new impressive large and late flowering *G. nivalis* introduction. Leaves applanate, slender, erect, blue-green. Outer segments large, rounded to bluntly pointed apex, good merging green lines above apex shadowed on underside. Inner segments broad inverted 'U'-shaped mark above sinus. Anette Brymer, Aargau, Switzerland, who has one of the largest collections in Switzerland. 2022. Vigorous. March

'Burley Bear'
Beautiful pure white *G. elwesii* Poculiform snowdrop. Leaves supervolute, erect to arching, broad, green-grey. Segments all of equal length, slender, pointed at apex, lightly longitudinally ridged. Cal Mateer, British Columbia.

'bursanus'
This distinct new snowdrop species was first described by Dimitri Zubov, Yildiz Konka and Aaron P. Davies when two small populations were found growing on rocky outcrops in the dry, oak forests of Bursana province, North-western Turkey. The snowdrop was named as a new species in 2019. Bulbs generally produce two scapes. Leaves are applanate to narrowly explicative, with two noticeable longitudinal folds which flatten towards the tip, small or absent at flowering, slender, strap-shaped, dark green at the base, both surfaces are glaucous with a noticeable central stripe and leaves twist slightly as plants mature. Outer segments, three, slender, rounded, lightly longitudinally ridged. Inner segments, three with either two green marks, one above the sinus and a second towards the base, or both marks can fuse forming an 'X' shape with a white central groove, the sinus mark forms a narrow green inverted 'V' shape. Flowers which appear in Autumn, usually around September, are highly fragrant. Due to its limited area and rarity *G. bursanus* is classed as an endangered species.
Autumn. 18-25cm.

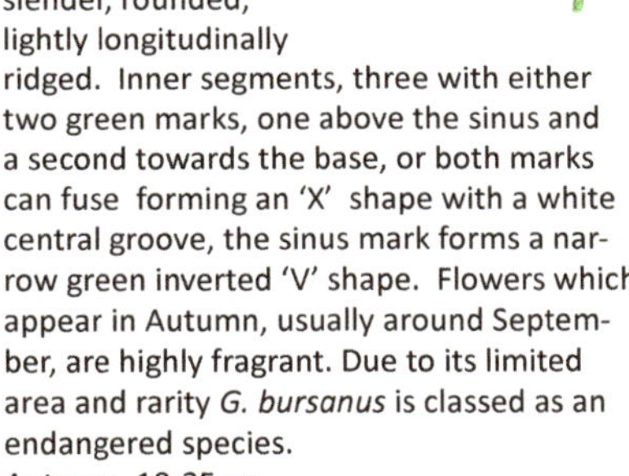

***'bursanus* Green Tip'**
Shorter growing strong-looking *G. bursanus*. Green spathe, pedicel and ovary. Leaves absent at flowering. Outer segments rounded to rounded apex, small light-green smudge above apex. Inner segments good strong green mark above sinus to two-thirds of segment.
Autumn. 10-15cm.

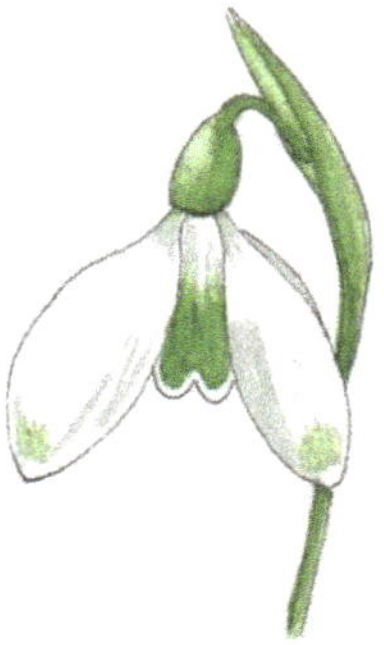

'Buttons and Bows'
Well-rounded flowers beneath dark-green rounded ovary and erect scapes. Outer segments broad, rounded, short claw, bluntly pointed at apex. Inner segments broad, rounded with double dark-green mark, narrow inverted 'V' above sinus beneath filled in 'V'-shaped mark towards base. Winter.

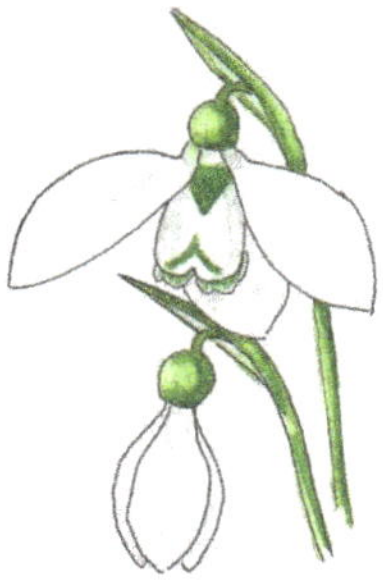

'Cal-u-Met'
Small flowered snowdrop with rounded, heavily green marked flowers beneath oblong dark green ovary. Outer segments broad, rounded, longitudinally ridged, slightly reflexed at apex, broad green mark above apex, second mark towards base. Inner segments slightly shorter than outer, longitudinally ridged, broad, green mark above apex, second mark towards base. Cal Mateer, British Columbia and named to commemorate the First Nation's Peace Pipes in memory of his grandmother. January.

'Calabrian Green'
Early *G. reginae-olgae* with long, slender green marked flowers. Outer segments long, slender, tapering to pointed apex, light-green marks above apex to two thirds of segment. Inner segments mid-green mark above sinus almost to base which can be divided longitudinally. Autumn/Winter.

'Cape Cod'
G.elwesii with well rounded flowers beneath oblong, rounded green ovary. Leaves supervolute, arching, broad, grey-green. Outer segments short claw, light goffering at base, rounded to rounded apex. Inner segments green inverted 'V'-shaped mark above sinus, shaded second mark towards base. USA introduction. Winter/Spring.

'Captain Dunstan'
Taller snowdrop with well-rounded buds and rounded, reflexed flowers. Leaves broad, erect to arching, green. Outer segments short claw, broad, incurved, rounded to pointed apex. Inner segments broad inverted green 'U'-shaped mark above sinus to half of segment.
Winter/Spring. 18cm.

'Captain Hook'
Unusual-looking, very random slender flowers on erect scapes with small, light-green ovary and pedicel. Leaves slender, erect to splayed, blue-green, lighter median line. Outer segments slender, incurved to 'hooked' apex. Inner segments rounded, variously marked green above apex.

'Carinthian Green Giant'
Heavily green-marked flowers on erect scapes with oblong, rounded green ovary. Outer segments rounded, tapering to pointed apex, merging green lines above apex towards base. Inner segments mid-green mark above sinus across segment almost to base, narrow white margin.
Gerhard Raschun, Austria.

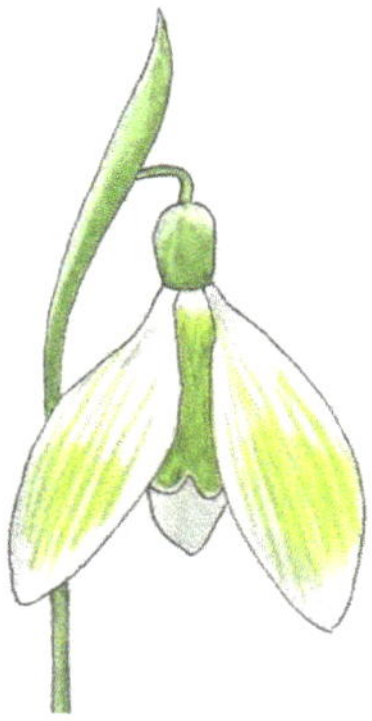

'Carlos aus Santana'
G. nivalis x G. elwesii with elegant, attractive flowers. Leaves broad, erect to arching, green. Outer segments incurved, rounded, tapering to bluntly pointed apex. Inner segments tinged cream, lightly longitudinally ridged, green heart-shaped mark above sinus, slightly running up ridges towards base, two paler ovals towards base.

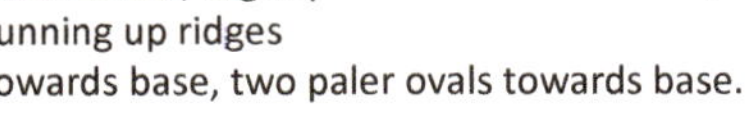

'Carol Dancer'
G. elwesii Inverse Poculiform snowdrop with good green marks. Leaves supervolute, broad, erect to arching, grey-green. Outer segments rounded to pinched, pointed apex, lightly longitudinally ridged, green mark above apex. Inner segments green mark above apex almost to base, narrow white margin. Named for one of the people in the group that found the snowdrop. Cal Mateer, British Columbia. Winter/Spring. 2021.

'Castle Green'
Shapely green-marked flowers with oblong green ovary. Leaves supervolute, erect, slender, blue-green. Outer segments rounded to bluntly pointed apex, light-green mark above apex almost to base. Inner segments green inverted 'U'-shaped mark above sinus, two small ovals towards base. Ian Christie.

'Castle Green Shadow'
Good Virescent snowdrop with oblong green ovary. Leaves slender, erect to arching, blue-green. Outer segments rounded to bluntly pointed apex, green lines above apex towards base. Inner segments dark green mark across segment from above sinus diffusing at base. narrow white margin. Ian Christie. Winter/Spring. 15cm.

'Castle Little Star'
Snowdrop with pretty, smaller flowers. Ian Christie. (No illustration)

'Castle New'
Rounded flowers on erect scapes with slender, angled pedicel and oblong rounded green ovary. Outer segments rounded, tapering to pointed apex. Inner segments, small narrow green inverted 'V'-shaped mark above sinus. Ian Christie

'Castle The Maules'
Rounded flowers with good green mark and rounded green ovary. Outer segments rounded to bluntly pointed apex, lightly longitudinally ridged. Inner segments green mark above sinus bleeding centrally towards base. Ian Christie. 2020. Winter/Spring. 16cm.

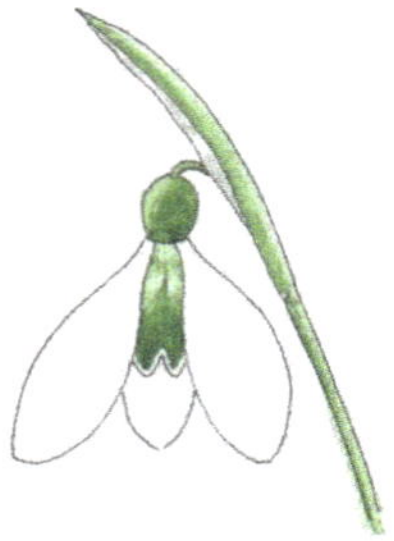

'Caterpillar'
Small, rounded, incurved, heavily green-marked flowers with broad spathe. Segments slender, rounded, variably marked green above apex to base. 10cm.

'Catherine McCauley'
A large, late flowering *G. plicatus* with well rounded flowers, broad segments and slender ovary. Leaves plicate, erect, grey-green, paler median line. Outer segments broad, rounded, bluntly pointed at apex. Inner segments mid-green mark above small sinus diffusing towards base. Ireland. 2014. January/March. 18cm.

'Celia Meggers'
Later *G. nivalis* with cream tinged flowers. Leaves erect, slender, grey-green. Outer segments rounded to pointed apex. Inner segments narrow green inverted 'V'-shaped mark above sinus. Nicholas Top and named for his sister having found it in the garden of the late Barbara Fluche in Germany. Vigorous. Spring. 15-16cm.

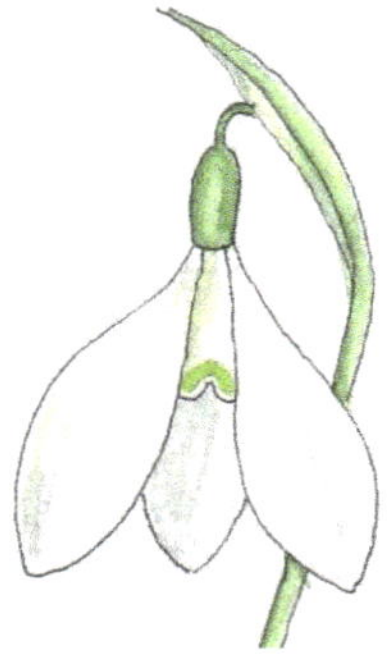

'Celia's Double'
G. x valentinei hybrid snowdrop with very rounded flowers on short spathe giving more outward facing flowers. Outer segments rounded towards pointed apex, lightly longitudinally ridged. Inner segments, tight neat ruff with green marks above sinus. Regularly two scapes one taller than the other. Named for Celia Sawyer when the seedling was found in her garden growing together with *G. plicatus* 'Diggory'. Very vigorous. Richard Bashford and Valerie Bexley, Woodchippings. 2016.

'Chadwick's Cream'
Smaller, rarer, distinctively cream coloured flowered *G. nivalis*, particularly in bud. Small rounded green ovary fading to yellow where it joins the pedicel. Leaves applanate, erect, slender, blue-green. Outer segments broad, rounded to bluntly pointed apex, tinged cream. Inner segments narrow, pale yellow-green inverted 'V'-shaped mark above sinus. Slow to bulk up and can be difficult to grow hence rarity. Nigel Chadwick. 10cm.

'Chantry Bonus'
Large, very substantial flowers. Leaves broad, grey-green. Outer segments large, long, up to 15mm, bluntly pointed at apex, longitudinally ridged, short green stripes above apex. Inner segments green mark across segment above sinus to base, narrow white margin. Wol and Sue Staines, Glen Chantry.

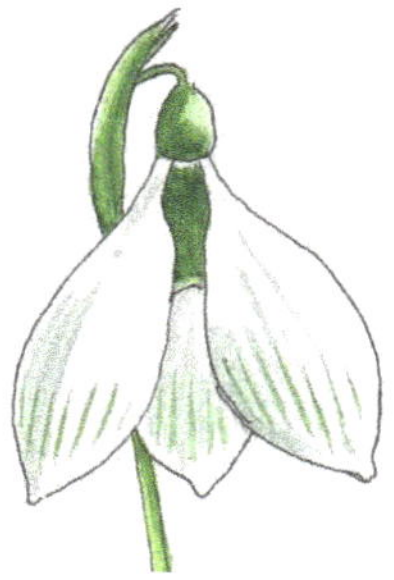

'Chantry Constable' Syn 'GC49'
Well-shaped flowers with good contrasting green marks and dark-green ovary. Outer segments rounded to bluntly pointed apex, bold, mid-green merging lines above apex to half of segment. Inner segments very dark-green mark above sinus almost to base, narrow white margin. Wol and Sue Staines, Glen Chantry. 2022.

''Chantry Cracker' - Syn. 'GC35
True to its name a 'cracking' snowdrop with well-shaped, green-marked flowers. Outer segments, short claw, broad, rounded to pointed apex, lightly longitudinally ridged, mid-green mark above apex bleeding into ridges. Inner segment good mid-green heart-shaped mark above sinus. Wol and Sue Staines, Glen Chantry. 2022.

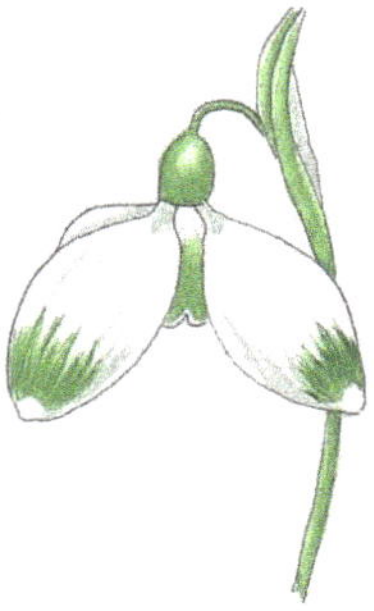

'Chantry Dame'
Good green marked flowers on erect scapes. Outer segments broad, rounded to rounded apex, longitudinally ridged, green mark above apex. Inner segment mark clearly visible even when flower closed, spreading across segment, apex to base, narrow white margin. Wol and Sue Staines, Glen Chantry.

'Chantry Dougal'
Green-marked Inverse Poculiform flowers with oblong green ovary. Outer segments broad, rounded, broadly rounded at apex, green mark above apex on lower third of segment. Inner segments slightly shorter with the inner markings said to resemble the face of Dougal in the Magic Roundabout children's series, the outer segments forming the droopy ears. Wol and Sue Staines, Glen Chantry. 2020.

'Chantry Duchess'
Imposing, large and elegant Inverse Poculiform snowdrop. Leaves broad, erect, grey-green. Six segments all of equal length, broad, flattened, rounded at apex, lightly ridged with green crescent-shaped mark above apex and paler green mark at base. Found as a seedling near *G.* 'E. A. Bowles'.
Wol and Sue Staines, Glen Chantry

'Chantry Empress'
Large, shapely flowers up to 15mm on tall, erect scapes with small dark-green ovary. Outer segments, long, slender claw, rounded to bluntly pointed apex, lightly longitudinally ridged and textured. Small green lines above apex. Inner segments broad green heart-shaped mark above sinus. Wol and Sue Staines, Glen Chantry.

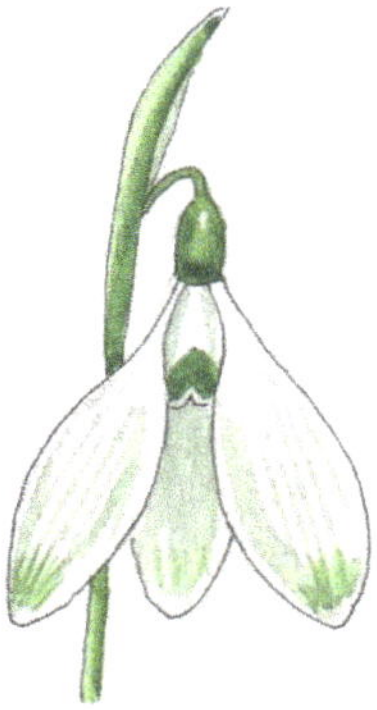

'Chantry Gold Cross'
A good, yellow marked *G.plicatus* suspended from slender yellow pedicel with slender yellow ovary. Outer segments long, slender, tapering to pointed apex. Inner segments good yellow 'V'-shaped mark above sinus merging into two yellow arms towards base creating a golden cross shape. Wol and Sue Staines, Glen Chantry. 2020. Winter/Spring. 15cm.

'Chantry Lady'
Very attractive, beautifully shaped Inverse Poculiform snowdrop suspended from slender pedicel with oblong green ovary. Outer segments broad, rounded, rounded at apex with edges flaring upwards, broad green spear-shaped mark to half of segment with slender line on basal side. Inner segments have a roughly 'H'-shaped mark. This snowdrop maintains its attractive shape throughout flowering. Wol and Sue Staines, Glen Chantry 2020.

'Chantry Lady Like'
A lighter green than G. 'Chantry Lady' with a more open flower. Outer segments rounded, rounded to apex, longitudinally ridged, broad green mark above apex. Long inner segments, green mark above sinus so half of segment bleeding into lines towards base.
Wol and Sue Staines, Glen Chantry.

'Chantry Pinstripe'
Small rounded triangular ovary. Outer segments rounded towards bluntly pointed apex, lightly longitudinally ridged, strong green lines above apex to half of segment, reminiscent of a pin striped suit. Inner segments green mark above sinus almost to base with small indent at base, possibly a 'tie' to go with the pinstripe suit? Wol and Sue Staines, Glen Chantry.

'Chantry Poppet'
Rounded green marked flowers with oblong rounded ovary. Outer segments broad, goffered at base, rounded to bluntly pointed apex, lightly longitudinally ridged with green lines from above apex on lower third of segment. Inner segments broad green mark above sinus, diffusing towards base.
Wol and Sue Staines, Glen Chantry.

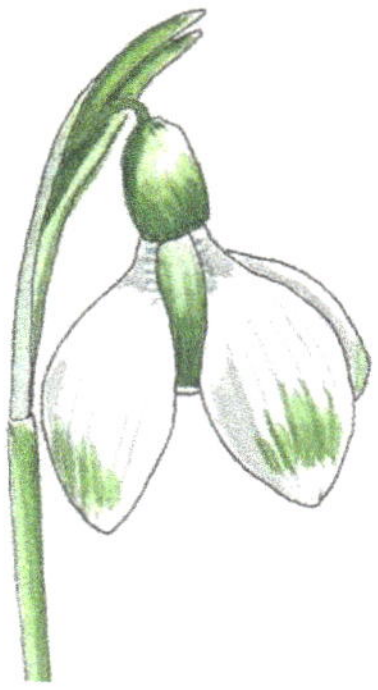

'Charley Barr'
Green-tipped *G x valentinei* snowdrop named for a landlord of the Pot Kiln pub, Yattenden, Paul Barney, Edulis Nursery. (No illustration)

'Charlie's Angel'
Slender, somewhat pointed flowers with rounded ovary. Leaves broad, erect, green. Outer segments clawed, incurved, tapering to pointed apex. Inner segments dark-green narrow inverted 'V'-shaped mark above sinus barely joining an oval towards base.

'Charmer-Flore-Pleno'
Delicate-looking *G. nivalis* with loose, open, green-marked double flowers. Leaves applanate, erect to splayed, slender, green. Outer segments four-five, slender, flaring, bluntly pointed at apex, light-green mark above apex. Inner segments loose ruff with light-green heart-shaped mark above sinus. Winter/Spring. 15cm.

'Che Bel Fior'
Heavily green-marked *G nivalis* with oblong green ovary. Leaves slender, erect to arching, applanate, blue-green. Outer segments rounded to bluntly pointed apex, Strong merging green lines from apex to three quarters of segment. Inner segments squared mid-green inverted 'U'-shaped mark above sinus to half of segment. Winter/Spring. 2020. 12cm.

'Chicken Pen'
Good, strong *G. elwesii* originating in Bill Baker's garden. No illustration.

'Child of Forest'
G. nivalis with shapely, green-marked flowers and oblong green ovary. Leaves applanate, erect to arching, slender, blue-green. Outer segments long, rounded to slightly pinched apex, lightly longitudinally ridged with strong green lines above apex, running up ridges to half of segment. Inner segments deep inverted mid-green 'U'-shaped mark above sinus merging into paler mark towards base. Snowdropfevers introduction. 2021. Winter/Spring. 16cm.

'Chiltern Queens'
A short, neat double *G. elwesii* seedling with tightly packed, well-rounded flowers on short scapes with small triangular green ovary. Outer segments broad, rounded, bluntly pointed at apex, occasional green mark at apex. Inner segments tight neat ruff of green marked segments, no sinus. Winter/Spring. 8cm.

'Chimaera'
Yellow marked *G. elwesii* snowdrop with attractive variegated foliage. Leaves supervolute, erect to arching, broad, variegated yellow-green. Outer segments rounded to bluntly pointed apex. Inner segments yellow mark above sinus.

'Chorus Line'
Syn 'NG 125' Erect scapes with rounded, oblong green ovary. Outer segments, clawed, rounded tapering to bluntly pointed apex. Inner segments double green mark, spreading inverted 'V'-shaped mark with rounded ends above sinus and a crescent-shaped mark towards base typically darker towards sinus. North Green Snowdrops.

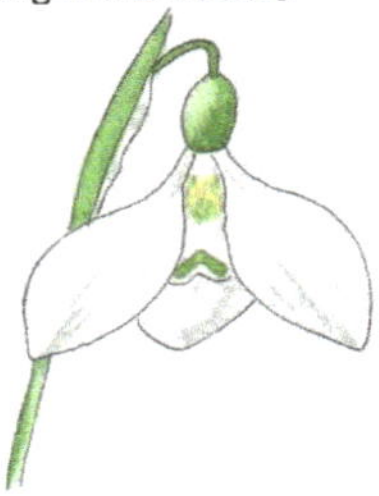

'Chthonic'
Not quite an albino *G. nivalis* but very nearly. Leaves applanate, slender, erect to arching, very distinctive pale creamy-grey. Outer segments slender, tapering to pointed apex. Inner segments small inverted pale-green-yellow inverted 'V' shaped mark above sinus barely joining on the basal edge. 'Chthonic' means emerging from the earth in the Underworld. Avon Bulbs, Somerset. 14cm.

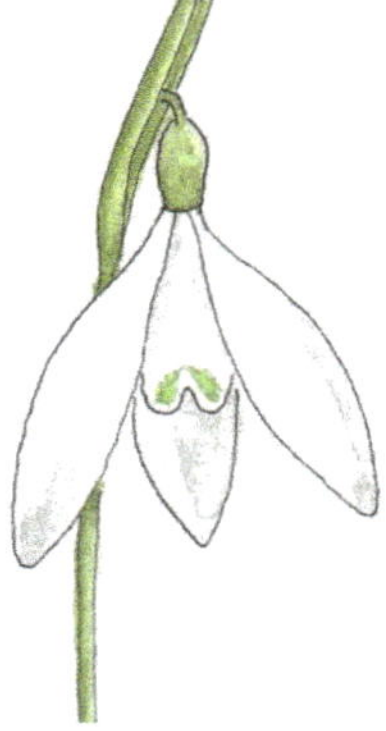

'Chthonic' means emerging from the earth in the Underworld. Avon Bulbs, Somerset. 14cm.

'Chubby'
A rather ambiguous-seeming name for an autumn flowering *G. reginae-olgae* with long slender flowers, erect scapes and small, rounded, dark green ovary. Outer segments long, slender, short slender claw, bluntly pointed at apex. Inner segments pale-green inverted 'V'-shaped mark above large sinus. An introduction from the USA. 12cm.

'Clemens Heidger'
G. plicatus subs. *byzantinus* with heavily green marked flowers. Outer segments rounded tapering to pointed apex, strong green lines above apex almost to base. Inner segments green mark above sinus across segment almost to base, narrow white margin. Hagen Engelmann, Germany and named for a friend.

'Clun'
Generally very reliable if not a spectacular *G. nivalis*. Leaves applanate, erect, slender, blue-green. Outer segments slender, pointed at apex. Inner segments broad inverted green 'U'-shaped mark above sinus. An early introduction *G. nivalis* by the Rev. R. J. Blakeway-Phillips c. 1980s. Some consider it the same as G. 'Clun Green' but others consider it is not necessarily the same snowdrop. Named for the village he lived in. Bulks up well. January/February. 14cm.

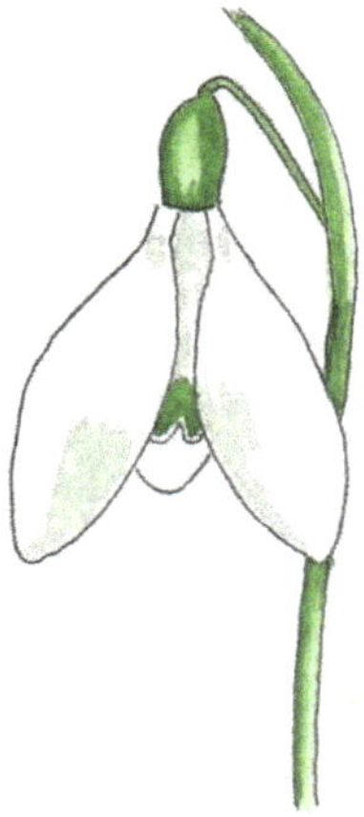

'Clun Green Convolute'
Small, well-rounded flowers. Leaves supervolute, erect, very narrow. Outer segments rounded, tapering to bluntly pointed apex. Inner segments solid green, slightly waisted mark above sinus across three quarters of segment, Narrow white margin. Rev. R. J. Blakeway-Phillips, Clun, Shropshire. c.1988.10cm.

'Cockapoo'
Unusual snowdrop with very variable flowers, throwing extra segments from the ovary and having a grumpy looking face. Paul Barney, Edulis Nursery, 2021.

'Cocoon'
Good strong growing *G. nivalis* with well rounded flowers. Leaves applanate, erect, slender, blue-green-glaucous. Outer segments broad, rounded, bluntly pointed at apex, short green lines above apex. Inner segments dark green inverted 'V'-shaped mark above sinus. Slow to bulk up. Winter/Spring. 12cm.

'Comet's Tail'
G. elwesii with broad, rounded flowers suspended from slender pedicel and oblong green ovary, throwing an extra narrow segment from the ovary. Leaves supervolute, broad, grey-green. Outer segments broad, flattened, tapering to bluntly pointed apex. Inner segments broad green mark above sinus.

'Compton Green Eye'
Striking snowdrop with slender flowers suspended from tall scapes. Outer segments slender, rounded to pointed apex. Inner segments light-green shadow mark. *G. elwesii x G. nivalis*. Paul Barney, Edulis Nursery. 2009. 23cm.

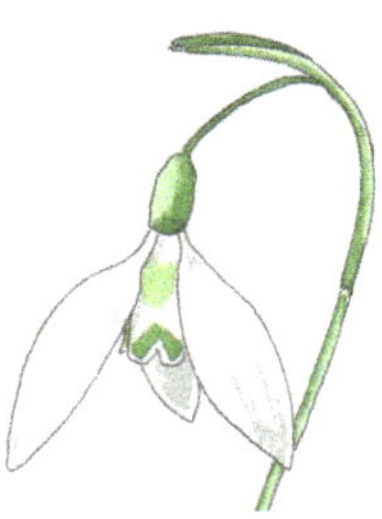

'Constellation'
G. nivalis with slender, green tipped, Poculiform flowers. Leaves applanate, erect to arching, slender, blue-green. Segments of equal length, slender, tapering to pointed apex, narrow green mark above apex.

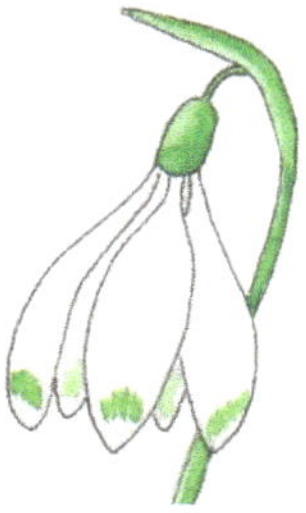

'Corovode'
G. reginae-olgae subs. *vernalis* from Albanian seed. Paul Barney, Edulis Nursery. 2020. January. (No illustration)

'Cowichan Caribou'
Small, stocky-looking *G. elwesii* Inverse Poculiform snowdrop with short scapes emerging from short, tightly clasping, blue-green leaves. Outer segments broad, rounded, lightly longitudinally ridged, rounded at apex, 'W'-shaped light-green mark above small sinus and two merging ovals towards base. Winter.

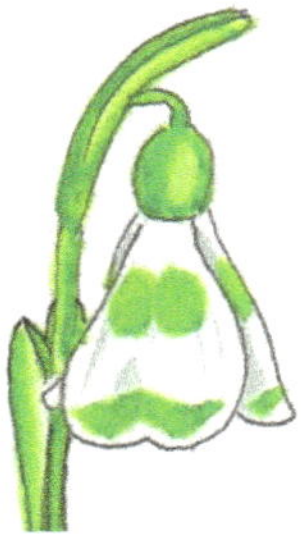

'Cream Bear'
Poculiform snowdrop with large well-rounded flowers starting cream in bud and fading as plants mature. Outer segments slightly larger than inner, well rounded and rounded at apex. Cal Mateer. British Columbia. 2022

'Cressida'
Elegant and beautifully shaped flowers suspended beneath slender green ovary. Outer segments rounded tapering to pointed apex. Inner segments narrow green 'V'- to 'W'-shaped mark above sinus, second green mark towards base.

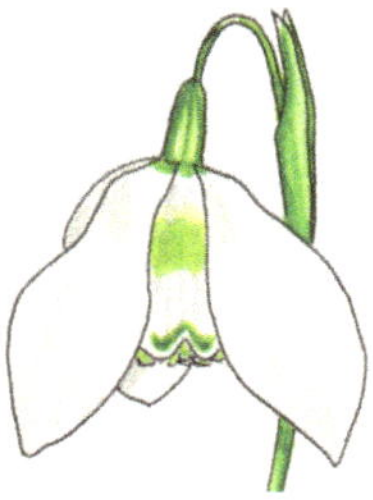

'Crime Wild'
Clean, bright-looking *G. plicatus* with arching pedicel and oblong ovary. Leaves explicative, slender, erect, blue-green. Outer segments short claw, rounded to bluntly pointed apex. Inner segments inverted heart shaped mark above sinus. January.

'Cross of Lattera'
Elegant-looking green marked flowers with oblong, rounded ovary. Outer segments slender, rounded into rounded apex, green lines above apex, washed green. Inner segments good green heart-shaped mark above sinus.

'Crown Jewel'
Beautifully shaped *G. elwesii* var. *monostictus* with triangular green ovary. Leaves supervolute, broad, grey-green. Outer segments spoon shaped, rounded at apex, lightly longitudinally ridged and textured. Inner segments broad, longitudinally ridged, small green dot either side of tiny sinus, two paler merging ovals midway across segments. Winter/Spring. 2022

'Cup of Joy'
Good new *G. nivalis* seedling with heavily green-marked flowers. Leaves applanate, erect to arching, slender, blue-green. Outer segments broad, rounded to apex, wide, merging green stripes above apex towards base. Inner segments dark-green mark above sinus bleeding towards base. 2022.

'Curse of the Were Rabbit'
A really curious and very variable *G. elwesii* in which every flower is different, displaying a variety of segments including petaloid spathes. Joe Sharman, Monksilver Nursery, Cambridge.

'Curtsey'
Unusual snowdrop with slender, flaring segments and oblong green ovary. Outer segments, four, tapering to pointed apex, narrow green line from above apex almost to base. Inner segments slender with broad green mark above sinus to three quarters of segment.

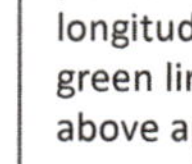

'Daft Punk'
A quirky *G. nivalis* snowdrop with extra segments flaring from a triangular green ovary. Leaves applanate, erect to arching, blue-green. Outer segments variable, slender, tapering to bluntly pointed apex, lightly longitudinally ridged. Inner segments broad, lightly longitudinally ridged with spreading inverted 'V'-shaped mark above large sinus. Edulis Nursery 2016.

'Dalkleid'
Poculiform snowdrop with slender dark-green spathe and triangular rounded ovary. Leaves medium, splayed, green. All segments roughly of equal length. Outer segments slender, rounded to bluntly pointed apex. Inner segments equal outer with small green mark above apex. Winter/Spring.

'Dangling Green'
Virescent *G. nivalis* with elongated flowers, good green marks and long slender green ovary. Leaves applanate, erect to splayed, slender, blue-green. Outer segments long, rounded to bluntly pointed apex, lightly longitudinally ridged, green lines from above apex to half of segment. Inner segments, broad green mark above sinus tapering to base. Winter/Spring. 16cm.

'Danish Gold'
Small, very slender pointed flowers with erect, slender pedicel and slender, oblong green ovary. Outer segments very slender, incurved to pointed apex. Inner segments small yellow mark above sinus.

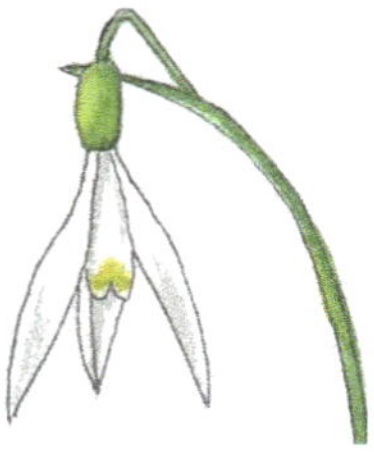

'David Foreman'
Very elegant, pagoda-shaped, long lasting, Inverse Poculiform flowers on erect scapes with oblong green ovary. Leaves erect to arching, grey-green, paler median line. Outer segments broad, flattened, rounded to rounded apex, lightly longitudinally ridged, yellow-green mark above apex and second yellow-green oval towards base. Inner segments mid-green mark above sinus almost to base, narrow white margin. Richard Bashford 2021. Winter/Spring. 6cm.

'Divine Comedy'
'Divine Comedy' - Green-tipped G. nivalis. 2020. (No illustration)

'Dizzy'
Unusually shaped flowers with undulating segments. Leaves erect, slender, green, paler median line. Outer segments undulating, tapering to bluntly pointed apex. Inner segments slightly shorter than outer, narrow green 'W'-shaped mark above sinus. Winter/Spring. 15cm.

'Doesn't Must'
Autumn-flowering *G. elwesii* var. *monostictus* with rounded ovary and broad, rounded, lightly ridged outer segments and small green mark above apex. Inner segments heart-shaped green mark above sinus. Ruben Billiet.

'Domaine du Oojevaar'
Later-flowering *G. nivalis* well marked with green. Leaves applanate, splayed, slender blue-green, paler median line. Outer segments rounded to bluntly pointed apex, lightly longitudinally ridged, merging green lines running up ridges from above apex to half of segment. Inner segments dark-green inverted 'V'-shaped mark above sinus. Domaine du Oojevaar Nursery, Flanders, Belgium and named for the nursery having been found in their garden. 2021. Winter/Spring. 14cm.

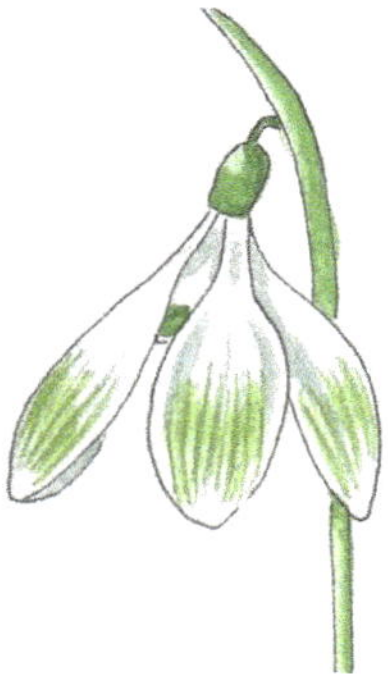

'Double Decker'
G. elwesii with well-rounded flowers on erect scapes beneath slender triangular ovary. Leaves supervolute, erect, broad, blue-green, incurved edges. Outer segments broad, rounded, rounded at apex. Inner segments good, mid-green heart-shaped mark above sinus. 2022. December.

'Double Dragon'
Quirky, flaring double flowers on this *G nivalis* which is almost a spiky. Leaves applanate, slender, erect, blue-green. Numerous outer segments giving a loose parasol effect, very slender, incurved, variable green marks above apex. Inner segments broader, with good heart-shaped mark above sinus. Ruben Billiet.

'Double Hedgehog'
Rare, short and neat snowdrop with upward facing flowers. Outer segments curving to bluntly pointed apex. Inner segments neat, green-marked ruff. Ian Christie.

'Double Orange Star'
Loose, open double flowers with flaring segments tapering to rounded apex. Inner segments variously marked green. Very pronounced orange anthers.

'Double Vision'
Large, well-textured flowers with oblong green ovary. Leaves broad, arching, ridged, hooded, blue-green. Outer segments clawed, broad, rounded, rounded at apex, longitudinally ridged and textured. Inner segments slender inverted green 'V'-shaped mark above sinus, shadowed immediately above before fading into two pale ovals towards base. 18cm.

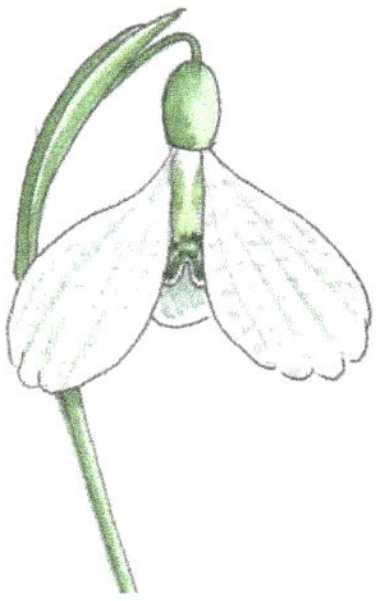

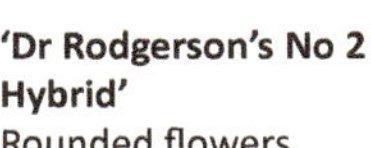

'Dr Rodgerson's No 2 Hybrid'
Rounded flowers beneath oblong green ovary. Leaves erect, slender, blue-green, paler median line. Outer segments rounded to bluntly pointed apex. Inner segments fused green mark above sinus, towards base.

Dragon Gold'
Attractive *G. nivalis* Sandersii with good yellow colouring, yellow green scape and spathe with well-shaped flowers suspended from oblong yellow ovary. Leaves applanate erect, slender, pale-green. Outer segments rounded to bluntly pointed apex. Inner segments slender, inverted light yellow-green 'V'-shaped mark with rounded ends above sinus. Belgium. 15cm.

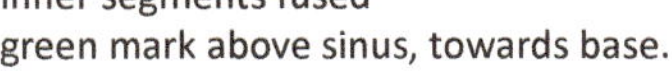

'Dryad Blizzard'
Superb Poculiform, the first in a new series of snowdrops, the 'snow' series from Dryad Nursery which have very pristine, white flowers. Tall, large, graceful, pure white flowers on erect scapes held well above leaves. Segments of equal length tapering to bluntly pointed apex, spreading and opening into a parasol shape. The long, slow progress of selecting this snowdrop began in 2008 when seed of a *G. elwesii* Poculiform snowdrop was received from Canada. These flowered in 2013 but sadly non were Poculiform. The seedlings were then inter-crossed and produced 6 seeds. The F2 seedlings flowered in 2017, giving one perfect Poculiform. This was assessed over the next five years giving a very tall, elegant, parasol shaped pure white Poculiform flower, finally released this year. Ann Wright, Dryad Nursery. 2022. 30cm.

'Dryad Cerberus'
Amusing small-flowered Inverse Poculiform snowdrop with three flowers from each spathe, and often a second scape from the same bulb, although these have reduced markings. Outer segments flattened, rounded to rounded apex with good sinus, broad, green inverted 'V'-shaped mark above sinus merging into a waisted oval towards base. Ann Wright, Dryad Nursery. Named for the dog with three heads. from the Underworld. 2019. Winter/Spring.

'Dryad Demeter'
Strongly marked Inverse Poculiform virescent snowdrop with slender-looking flowers forming a pyramid shape with slightly flaring segments. Leaves plicate, slender, erect to splayed, grey-green. Outer segments flattened, tapering to rounded apex with good sinus, broad mid-green mark above sinus almost to base, narrow white margin. Inner segments mid-green mark almost to base, above deep sinus, narrow white margin. Good perfume. Dryad Myths and Legends series and named for the goddess of plants and agriculture. Winter/Spring. 16-18cm

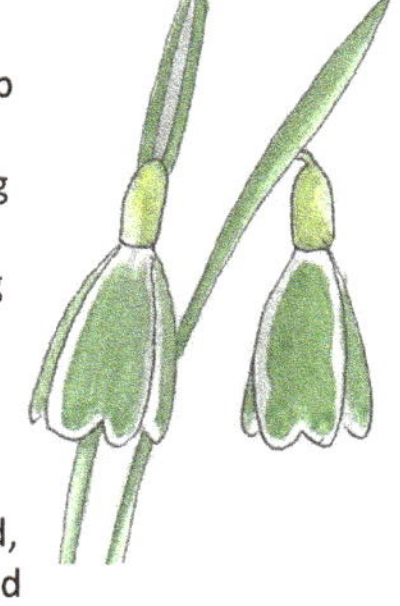

'Dryad Echo'
A tall snowdrop with large, imposing flowers and the green marks on outer and inner segments repeated above as a paler 'echo'. Leaves plicate, erect to arching, grey-green. Outer segments short claw, broad, flattened, rounded to rounded apex, longitudinally ridged, strong green 'U'-shaped mark above good sinus, paler mark shadowing above. Inner segments shorter than outer with narrow, green inverted 'V'-shaped mark above sinus and paler mark shadowed above almost to base. Good honey perfume. Vigorous and bulks up well. Named in the Dryad Myths and Legends series of snowdrops for Echo, spirit of the mountains and woods who was doomed to repeat the last words she heard. Introduced in 2020 after many years of work and evaluation. Ann Wright, Dryad Nursery. 2020. Winter/

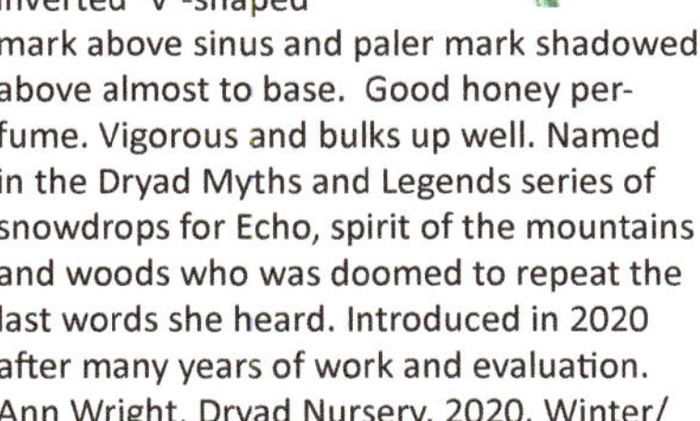

'Dryad Gold Sceptre'
Stately, attractive gold-marked flowers on erect scapes, arching pedicel and oblong, rounded yellow ovary. Leaves glaucous. Outer segments, short claw, tapering to pointed apex. Inner segments inverted broad, yellow, horseshoe-shaped mark above sinus. Large bulbs, bulks up well. Good honey perfume. Named for its erect stance from a gold staff or sceptre. Ann Wright, Dryad Nursery, 2020. Winter/Spring. 15-20cm.

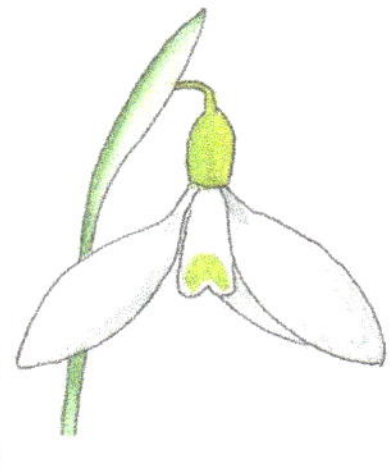

'Dryad Gold Standard'
Another snowdrop in the 'Gold' series from Dryad Nursery with large, bell-shaped flowers on tall, erect scapes, slender pedicel, and rounded-oblong yellow ovary, Leaves grey-green, glaucous, central ridge, erect and later arching. Outer segments incurved, short claw, pointed at apex. Inner segments inverted yellow heart-shaped mark above sinus. Large bulbs, bulks up well and quickly. In the Dryad Gold Series of snowdrops. Anne Wright, Dryad Nursery, 2020. Winter/Spring. 15-18cm.

'Dryad Hera'
Inverse Poculiform flowers similar marking to *G.* 'Demeter' but considered more 'feminine' looking. From Dryad's myths and legends series and named for Demeter's sister for being green as she was so jealous of her father/husband, Zeus's, liaisons with other women. Anne Wright, Dryad Nursery.

'Dryad Jupiter'
A large, imposing Inverse Poculiform from Dryad Nursery's Myths and Legends series of snowdrops. Strong, tall erect scapes, large flowers standing well above leaves. Outer segments short claw, broad, rounded, distinctly longitudinally ridged with upward facing raised edge to segments, rounded green mark above small sinus bleeding into ridges giving the appearance of a 'crown'. Large bulbs, bulks up well. Named for king of the Roman Gods. Ann Wright, Dryad Nursery. 2021.

'Dryad Terpsichore'
A tall, neat looking Inverse Poculiform snowdrop in the Dryad Myths and Legends series. Leaves erect, plicate, short at flowering. The flaring flowers seem to dance. Outer segments broad, rounded, flaring, lightly longitudinally ridged with green mark above the sinus. Inner segments good mid-green inverted heart-shaped mark above sinus. Well perfumed. Named for one of the muses in Greek legend and thought to have been the mother of the Sirens whose singing lured sailors to their deaths. Anne Wright, Dryad Nursery, 2022. 20cm.

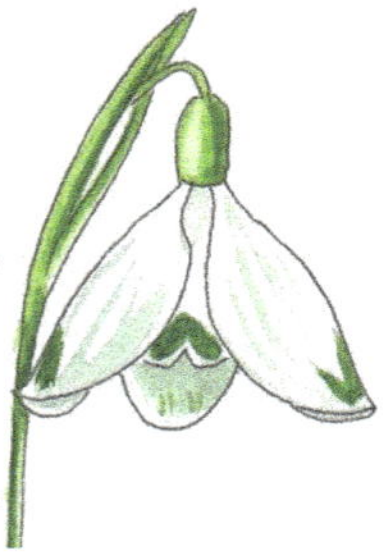

'Dryad Venus'
Attractive, well-shaped flowers with erect scapes and rounded ovary. Leaves plicate and spreading. Outer segments short claw, broad, flattened, longitudinally ridged, flaring at edges especially in mature flowers, broad green mark above apex often diffusing towards base. Inner segments slightly flared at edges with good sinus and large green 'X' shaped mark. The snowdrop is in Dryad Myths and Legends series and named for Venus, Goddess of love and beauty because of its attractive shape. Good perfume. Ann Wright, Dryad Nursery 2018. 18-20cm.

'Dryad Zeus'
Striking Inverse Poculiform flowers on erect scapes with erect pedicel, long slender green ovary and strong green segment marks. Leaves plicate and to around half of scape. Outer segments parallel sided long green strip down centre of segment, rounded towards apex with white notch, paling yellow-green towards base, mark can have a slight waist and also break into two marks. Inner segments green stripe above apex paling towards base. Parents *G.* 'Wendy's Gold *x G.* South Hayes. Dryad Nursery Myths and Legends series, named for the God of the sky and Thunder with its lightening bold green stripe down the centre of the segments. 2022. 18-20cm.

'DSC-3066 Tuuliku'
Well-shaped flowers with yellow-green spathe, pedicel and ovary. Outer segments rounded, tapering to pointed apex. Inner segments narrow yellow inverted 'U' to 'V'-shaped mark above sinus.

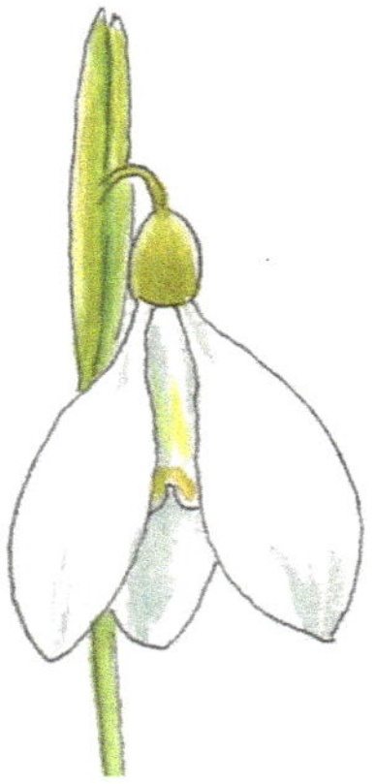

'Duizendduin'
G. elwesii x G. nivalis hybrid with elegant green marked flowers, light green oblong ovary and arching pedicel. Leaves erect, broad, green-grey. Outer segments clawed, lightly longitudinally ridged, green lines above apex to half of segment. Inner segments light green mark across segment from above sinus, almost to base, narrow white margin. Winter/Spring.

'Dumbo'
Later flowering snowdrop with spreading, broad, flattened segments and oblong, rounded green ovary. Leaves erect, slender, green-blue, paler median line. Outer segments broad, flattened, rounded at apex, green crescent-shaped mark above apex. Inner segments broad inverted green 'U'- shaped mark above sinus. Spring. 15cm.

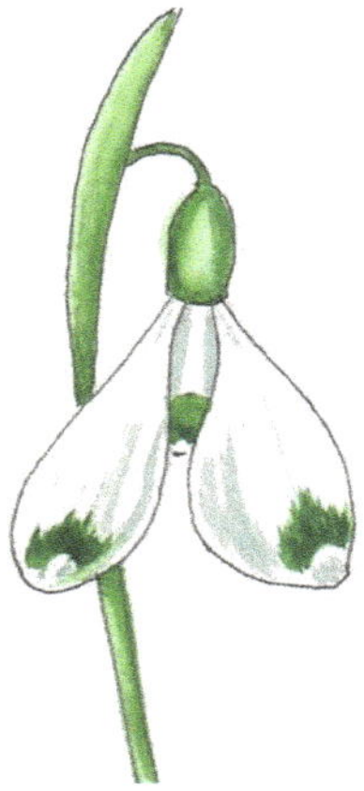

'Early Autumn Jewel'
Slender-looking *G. elwesii* flowers on erect scapes with small rounded green ovary. Leaves broad, erect to arching, grey-green. Outer segments slender, tapering to pointed apex. Inner segments green mark across segment from above sinus, divided longitudinally, narrow white margin. Autumn/ Winter.

'Early Green'
Later-flowering, heavily green-marked *G. nivalis*. Leaves applanate, erect to arching, slender, blue-green. Outer segments rounded to bluntly pointed apex, light goffering at base, green stripes above apex to three quarters of segment, pale-green wash to base. Inner segments mid-green mark across segment above sinus to base, narrow white margin. John Alpassa. Spring. 15cm.

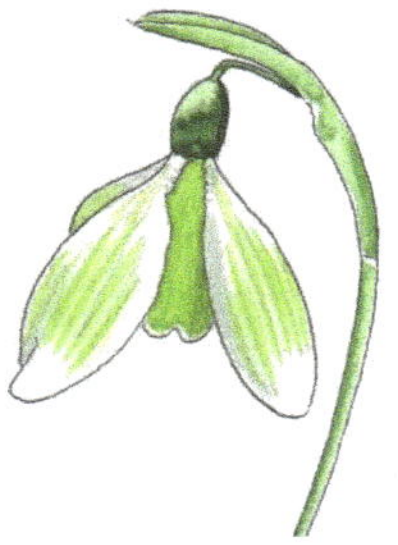

'Easter Parade'
Neat, open, taller *G. nivalis* 'Flore-Pleno' with large flowers on erect scapes and neat inner ruff. Leaves applanate, erect to arching, slender, green. Outer segments flaring, slender, tapering to pointed apex. Inner segments neat ruff with narrow green inverted 'V'-shaped mark above sinus. Vigorous, bulks up well and quickly. Found in Tipperary, Ireland on an Easter Sunday. Field of Blooms. 2019. 16cm.

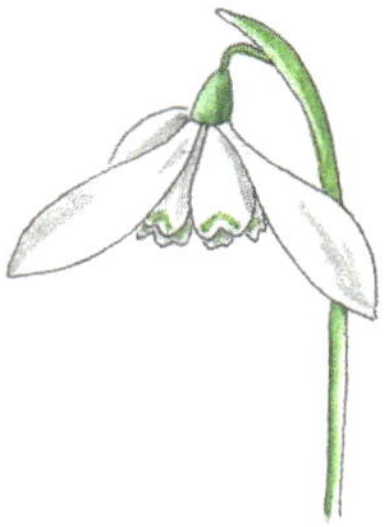

'Echoes'
Smaller *G. elwesii* often with two, well-shaped, large flowers suspended from a single pedicel. Leaves supervolute, broad, erect to arching, grey. Outer segments rounded tapering to bluntly pointed apex. Inner segments green mark above sinus. Winter/Spring. 12cm.

'Echo'
Similar-looking snowdrop to *G.* 'Robin Hood' with flowers suspended from slender pedicel on tall, erect scapes. Leaves slender, erect, blue-green-glaucous. Outer segments long, clawed, rounded, tapering to pointed apex. Inner segments green mark across segment from above sinus. Good, strong grower. Probably hybrid between *G. plicatus* and *G. elwesii*. Found by Bill Clarke, Wandlebury, Cambridgeshire and named for his grandson. Slow to increase. Winter. 20cm.

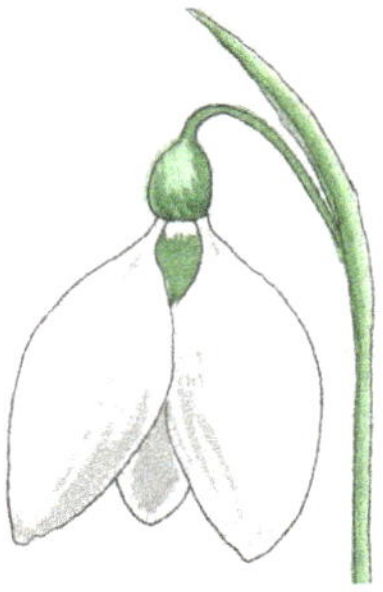

'Edward Ray Robie'
Unusual *G. elwesii* with wide spreading flowers, curving scape, twisted spathe and angled pedicel. Leaves supervolute, broad, spreading, grey-green. Outer segments spreading, rounded towards apex and bluntly pointed. Inner segments absent showing yellow anthers with a green line (similar to *G.* 'Yellow Star' which doesn't have the green line). Cal Mateer and named for someone special. Winter flowering.

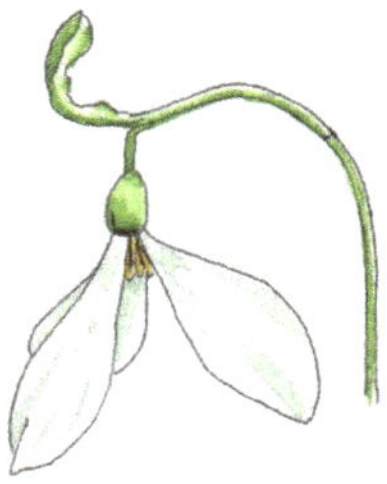

'Eeyore'
Large, rounded flowers on erect scapes with broad spathe, slender pedicel and rounded green ovary. Outer segments, broad, rounded, longitudinally ridged, light goffering at base, green mark above sinus. Inner segments long with broad inverted 'V'-shaped mark above deep sinus. Two small paler green ovals towards base. 18cm.

'Eifeler Grün'
Tall, erect scapes bearing long slender flowers. Leaves erect, slender, green. Outer segments long, incurved, green mark above apex extending in thin line towards base. Inner segments broad green mark above sinus. Rudi Bauer, Germany. 2012. 15cm.

'El Bandito'
G. elwesii with slender looking flowers and oblong green ovary. Leaves supervolute, erect to arching, broad, grey-green. Outer segments slender, tapering to pointed apex. Inner segments green inverted 'V'-shaped mark above sinus, two 'eye' spots towards base. Paul Barney, from Bill Baker's Garden. 2011.

'El Mono'
Ex-'Pine Knot' - *G.elwesii* with well-shaped flowers and long, slender, green ovary. Leaves supervolute, erect, medium, grey-green. Outer segments pronounced claw, rounded, rounded at apex. Inner segments broad green mark above sinus almost to base, narrow white margin. 15cm.

'Elegant Bear'
Well-shaped, elegant and distinctive Poculiform flowers with soft green marking. Outer segments rounded to pointed apex. Inner segments slightly shorter than outer with light-green marks above apex. Cal Mateer, British Columbia. 2021. Winter/Spring.

'Elizabeth Benfield'
Good *G.gracilis* hybrid with distinctive marks. Leaves applanate, erect, slender, very grey. Outer segments long, slender, incurved to bluntly pointed apex. Inner segments green mark either side of sinus and second paler more diffuse mark towards base. Very vigorous and bulks up well.

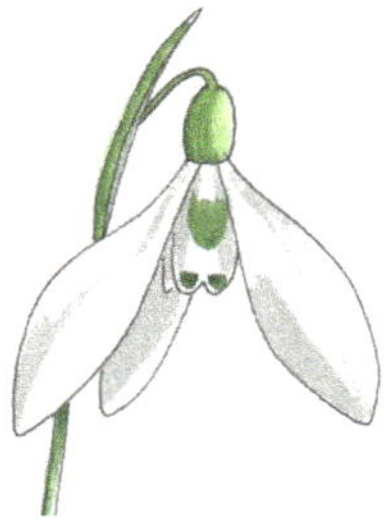

'Ellie Saunders'
Small, very neat and regular *G.plicatus* with Inverse Poculiform flowers. Leaves explicative, erect to arching, dark green with paler median line and broad margins. Outer segments flattened, rounded to bluntly pointed apex, strongly longitudinally ridged, dark-green mark above apex, paler mark towards base. Very vigorous. 10-15cm.

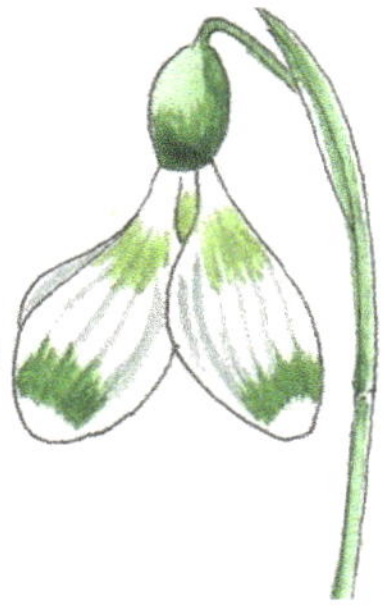

'Elven Lantern'
Small, strange looking, very slender Poculiform flowers. Leaves erect to arching slender, grey-green. Slender, green-marked segments roughly of equal length, incurved, rounded to bluntly pointed apex, marked green above apex. Orange anthers clearly showing at base. Winter/Spring. 12cm.

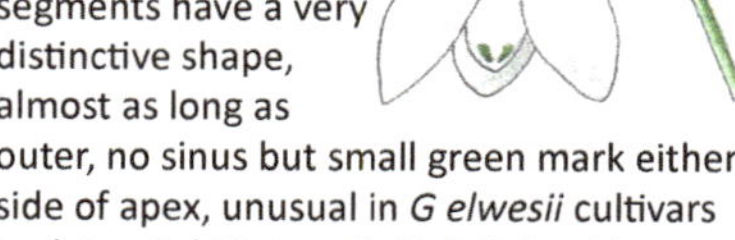

'Emerald Bear'
A semi-Poculiform *G. elwesii* with rounded flowers on slender pedicel with rounded green ovary. Leaves broad, spreading, grey-green. Outer segments broad, rounded to bluntly pointed apex. Inner segments have a very distinctive shape, almost as long as outer, no sinus but small green mark either side of apex, unusual in *G elwesii* cultivars to date. Cal Mateer, British Columbia.

'Emerald Pagoda'
Smaller snowdrop with erect scapes, oblong ovary and outward facing flowers forming a pagoda shape. Leaves erect, sage-green. Outer segments broad, flattened, rounded towards apex, lightly longitudinally ridged, green mark above apex. Autumn/Winter. 10cm.

'Eppie'
Large, well-rounded, textured flowers with oblong green ovary. Outer segments broad, rounded, rounded at apex, longitudinally ridged and textured. Inner segments green mark above sinus

'Equinox'
Very late and tall flowering *G. nivalis* helping extend the season Long green ovary. Leaves applanate, erect to arching, slender, blue-green. Outer segments slender, rounded, tapering to pointed apex. Inner segments green inverted tall 'V'- shaped mark above sinus joining second mark towards base.
Often in flower for the spring equinox. Bulks up well making good clumps. Normandy. 2009. Spring. 16-18cm.

'Eric Covell'
Early flowering *G. elwesii* var. *monostictus*. Leaves supervolute, broad, grey-green. Outer segments lightly longitudinally ridged tapering to bluntly pointed apex, small green marks above apex. Inner segments green mark from above sinus to half of segment. Edgwood Gardens, USA.

'Eric Saints'
Good green marked *G. elwesii* var. *monostictus* with well shaped flowers. Leaves supervolute, broad, grey-green. Outer segments broad, rounded to bluntly pointed apex, green mark above apex. Inner segments broad, rounded, broad green heart-shaped mark above sinus. Winter/Spring. 15cm.

'Eric Watson'
Compact *G. plicatus* snowdrop with large, rounded clean white flowers on slender pedicel with oblong, rounded, green ovary. Leaves plicate, erect, grey-green. Outer segments rounded tapering to bluntly pointed apex. Inner segments green inverted 'V'-shaped mark above sinus. From Eric Watson. Winter/Spring. 8cm.

'Ernie Cavallo'
An excellent *G. woronowii* with beautiful colouration and large semi-virescent flowers suspended from sturdy, erect scapes with oblong green ovary. Leaves strongly supervolute, erect to arching, green, often with third leaf present. Outer segments rounded to bluntly pointed apex, yellow green colour-wash, tipped white at apex and along borders. Inner segments broad green molar shaped mark above sinus fading into yellow green on basal side., distinctive narrow white margin. Patricia Becker, New Jersey, USA. 2018. Selected and named for an American Galanthohphile, making him a 'Snowdrop Immortal'. 12cm.

'Essex Girl' Syn. 'GC45"
Very distinctive Hybrid Inverse Poculiform with broad, pleated-edged segments and slender, oblong rounded dark-green ovary. Outer segments broad, flattened, frilled edges, rounded at apex, longitudinally ridged with crescent shaped green mark above apex. Inner segments dark green Chinese-bridge shaped mark above sinus and two narrow green marks at base from ovary. Vigorous. Wol and Sue Staines, Glen Chantry. 2019.

'Étoile des Neiges'
Quirky snowdrop that stands out in the crowd. Neither a true spiky or a double. Leaves applanate, slender, erect to splayed, blue-green. Segments numerous, slender, variable, arranged in a tightly packed cluster. Discovered growing in colonies of *G. nivalis* and *G. nivalis* 'Flore Pleno' by Trudi Veco-Desmet. 12cm.

'ex Cappoquin-House'
G. elwesii with dark-green oblong rounded ovary. Leaves supervolute, broad, arching, grey-green. Outer segments rounded to apex. Inner segments squared, dark-green inverted 'V'-shaped mark above sinus and two merging ovals towards base. Winter/Spring.16cm.

'ex Cherub Seedling'
Stunning *G. plicatus* with well-rounded, heavily crinkled, balloon-like flowers. Leaves explicative, erect, broad, grey-green. Outer segments very broad, rounded to rounded apex, strongly longitudinally ridged and textured, far more textured than *G.* 'Cherub'. Inner segments mid-green mark above sinus almost to base, narrow white margin. Richard Bashford. 16cm.

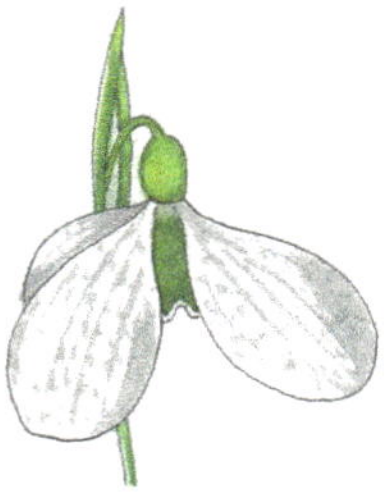

ex 'Gedelme'
Rounded flowers on this neat looking *G. peshmenii*. Leaves applanate, erect to arching, slender, grey-green. Outer segments short claw, well-rounded to bluntly pointed apex. Inner segments small green marks at sinus. Autumn/Winter.

ex 'Gert Geenson'
Slender flowers with long, angled pedicel. Leaves erect, slender, green. Outer segments slender, rounded, lightly longitudinally ridged, rounded to bluntly pointed apex, small green lines above apex. Inner segments tube-like, narrow green inverted 'V'-shaped mark above sinus.

ex 'Montrose'
G. elwesii var. *monostictus* with erect scapes and long slender ovary. Leaves, supervolute, broad, erect, green. Outer segments rounded tapering to bluntly pointed apex. Inner segments broad green inverted 'U'-shaped mark above sinus. Autumn/Winter. 15cm.

ex 'Ted Kiely'
Tall erect scapes bearing large, well-rounded flowers and oblong green ovary. Leaves broad, erect, recurving at tip, grey-green. Outer segments, broad, boat shaped, lightly textured tapering to bluntly pointed apex. Inner segments, inverted green 'V'-shaped mark above sinus, two merging ovals towards base. Winter/Spring. 15cm.

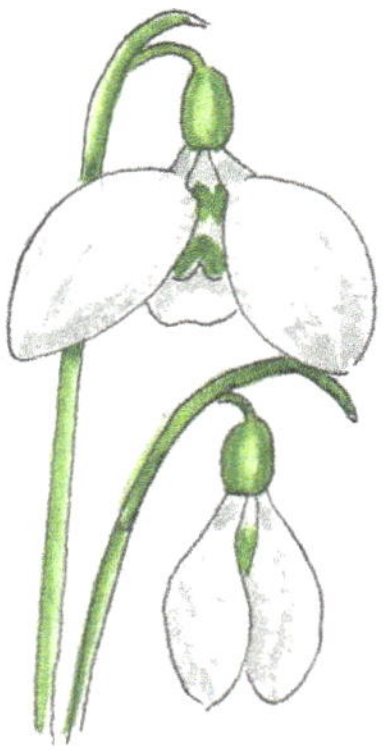

'Eye Eve'
Rounded flowers with rounded ovary suspended from slender pedicel. Outer segments clawed, rounded, tapering to pointed apex, lightly longitudinally ridged. Inner segments green mark above small sinus, second mark towards base.

'Fairlight'
Smaller later-flowering *G. elwesii* with slender looking flowers and rounded green ovary. Leaves supervolute, broad, erect, short, pointed and clasping. Outer segments clawed, small green mark above apex. Inner segments small green mark either side of shallow sinus, two paler yellow-green ovals towards base. Dorothy Underhill's garden, Collaton St Mary, Devon, and named for her house. Winter/Spring. 10cm.

'Fanfreluche'
G. nivalis spiky with multiple segments making for a large, fat, rounded flower with small green ovary. Leaves applanate, slender, splayed, blue-green. Outer segments slender, curving to bluntly pointed apex, small green lines above apex. Inner segments shorter than outer, green inverted 'V'-shaped mark above sinus. Oliver Vico, Domaine du Oojevaar, Flanders, Belgium.

'Fast Zerdatscht'
Bavarian *G. nivalis* Inverse Poculiform snowdrop with oblong green ovary. Leaves applanate, erect to splayed. slender, blue-green. Segments rounded to rounded apex, slight sinus, green mark above sinus. Horst Bauerlein, Bavaria. Translation means 'almost squashed' as he nearly trod on it when walking round his overgrown garden. 2019. 10cm.

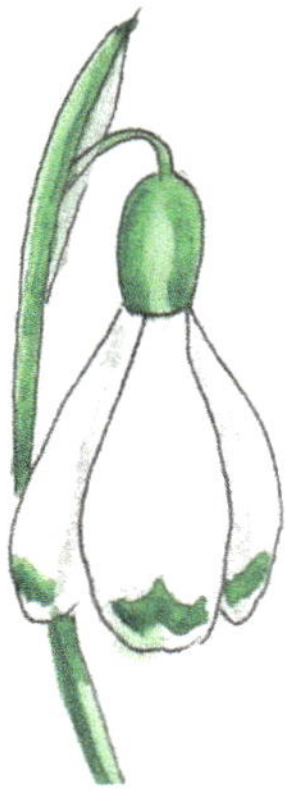

'Feline'
Smaller, well-shaped flowers on erect scapes. Flattened rounded segments with longitudinal ridging and small green mark either side of sinus. Could be an Inverse Poculiform or a Pociliform 2022.

'Fieldgate Fantasia'
Outstanding *x valentinei* Poculiform snowdrop with beautifully shaped, heavily green-marked flowers on erect scapes. Leaves slender, erect to arching, blue-green-glaucous. Outer segments broad, rounded, rounded at apex, longitudinally ridged, green mark above apex bleeding into ridges, second pale mark at base. Inner segments, green mark above sinus to base. Colin Mason.

'Fieldgate Strong'
Good green-marked flowers with rounded green ovary beneath broad green spathe. Outer segments long rounded at apex, green mark above apex and second mark at base. Inner segments green mark from above sinus to base, narrow white margin. Colin Mason. Winter/Spring.

'Fifi'
Small, slender, green marked flowers on long erect pedicel from base of spathe. Outer segments long, slender tapering to bluntly pointed apex, strong green inverted 'V'-shaped mark above apex. Inner segments to half of outer, green mark across segment, diffusing at base, narrow white margin.

'Fireside Find'
Good, strong growing *G. elwesii* Inverse Poculiform from British Columbia. Leaves supervolute, erect, broad, grey-green. Outer segments broad, rounded to rounded apex, green mark above apex and second mark towards base. Inner segments double green mark as on outer. Found near the Fireside Grill, Victoria. BC. by Carol Dancer. 2022. 10-15cm.

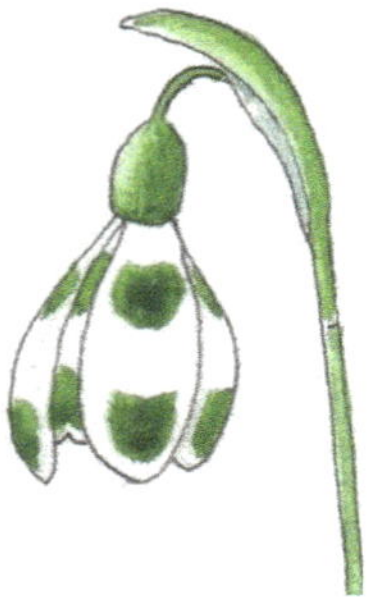

'Flame'
Inverse Poculiform with green-marked flowers. Leaves erect, broad, grey-green. Outer segments broad, flattened, rounded at apex, lightly longitudinally ridged, broad green mark above apex merging into yellow-green mark at base. Inner segments green mark from above sinus to base. Winter/Spring. 16cm.

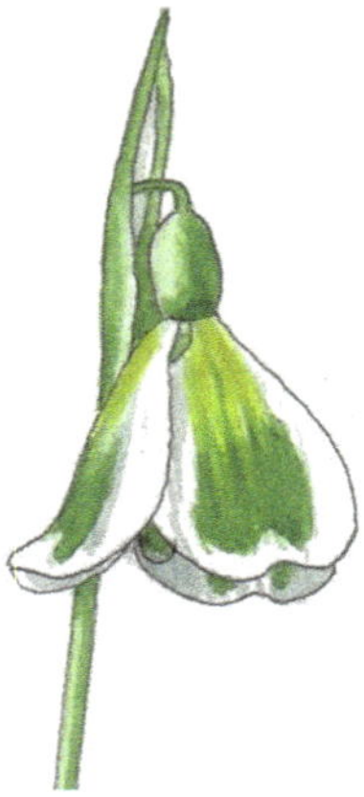

'Flanders Beauty'
Large bulbs producing slender, heavily green marked flowers on erect scapes with dark green ovary, scape and spathe. Outer segments slender tapering to bluntly pointed apex. Dark-green marks above apex almost to base, narrow white margin. Inner segments dark green mark above sinus diffusing towards base. Freddy Van Houtte Winter/Spring.

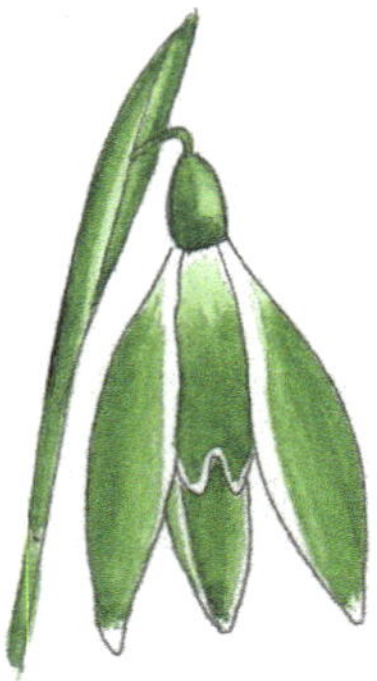

Flanders Field'
Beautiful, heavily green-marked, fully double flowers. Three flaring outer segments, rounded and incurved to apex, light green marks at apex. Inner segments neat ruff of heavily green-marked segments.

'Flavia'
Good combination of green and gold. Green spathe and scape, yellow pedicel and ovary. Leaves splayed, slender, green. Outer segments rounded to pointed apex. Inner segments pale lemon-yellow mark above sinus, paler mark at base. Winter/Spring. 8cm.

'Floriferous Star'
A new American seedling worth watching and possible could be re-named later if stable. An Inverse Poculiform snowdrop with spreading, cup-shaped flowers having four outer segments and extra inner segments. Outer segments rounded, spreading, lightly longitudinally ridged, merging green stripes inside and out above rounded apex. Inner segments marked green, revealing yellow stamens.

'Forest Muse'
Delicate looking *G. nivalis* with cream tinged buds. Leaves applanate, erect, slender, blue-green. Outer segments slender, rounded to pointed apex. Inner segments flushed yellow, green mark above sinus which changes to yellow as flowers mature. 2019. 10-15cm.

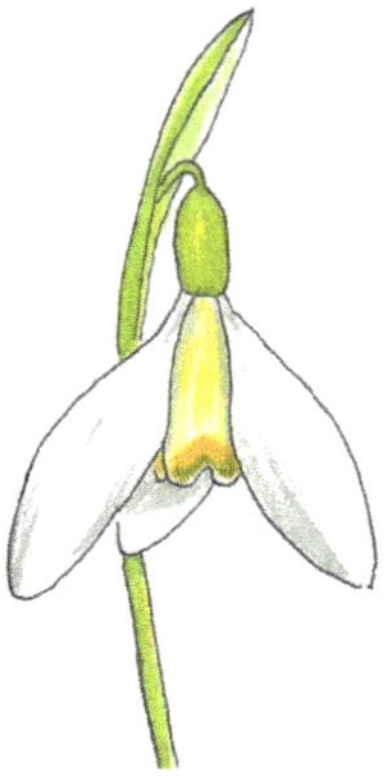

'Forger's Brush'
A chance seedling found in a clump of normal *G. nivalis*. Slender, rounded green-marked segments. Leaves erect to splayed, slender, blue-green. Outer segments short claw, rounded to bluntly pointed apex, soft green mark from above apex almost to base. Inner segments mid green mark above sinus paling towards base, narrow white margin. Nicholas Hecket and named as it resembles more expensive snowdrops, hence 'forger'.

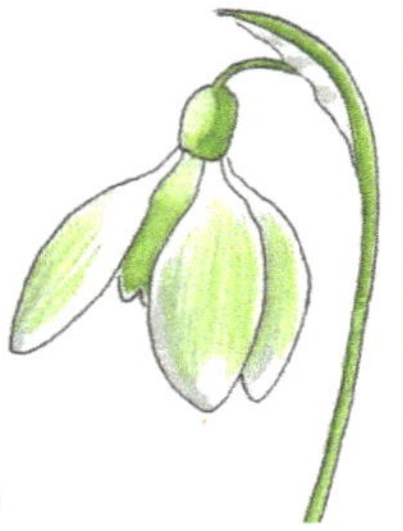

'Four Play'
Slender-looking *G. nivalis* with four outer segments and good green marks. Leaves applanate, erect to arching, slender, blue-green. Outer segments slender, tapering to pointed apex, green wash above apex. Inner segments mid-green heart-shaped mark above sinus. Winter/Spring. 15cm.

'Frail Beauty' Syn. Syn. NG 004
Erect blue-green leaves and upright spathe with slender pedicel and well marked, rounded flowers. Outer segments broad, rounded, broad shoulder, rounded towards apex, strong green lines from above apex towards base. Double inner segment marking clearly shows through very thin, finely textured flowers. Requires a sheltered position. North Green 2022. 8cm.

'Freak'
Quirky *G. nivalis* with erect scapes, slightly split spathe and long pedicel. Leaves applanate, slender, erect to arching, green-blue-glaucous. Ovary throws extra green marked segments from the top. Outer segments rounded, tapering to rounded apex. Inner segments broad, lightly longitudinally ridged, green spot either side of sinus. Winter/Spring.

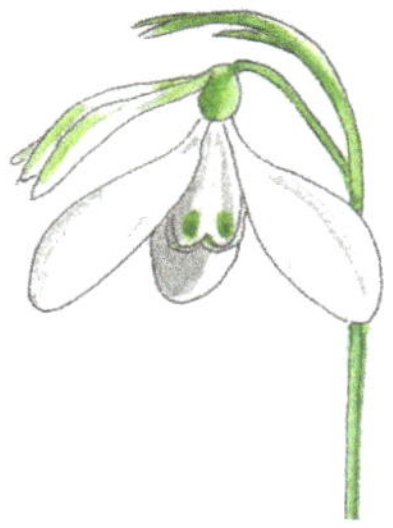

'Fred's Surprise'
Smaller snowdrop with long Scharlockii type split spathe, long slender pedicel, and small green-marked flowers held lower than leaves. Leave slender, erect, blue-green, paler median line. Segments roughly of equal length, heavily marked green above apex to half of segment. 2020. 10cm.

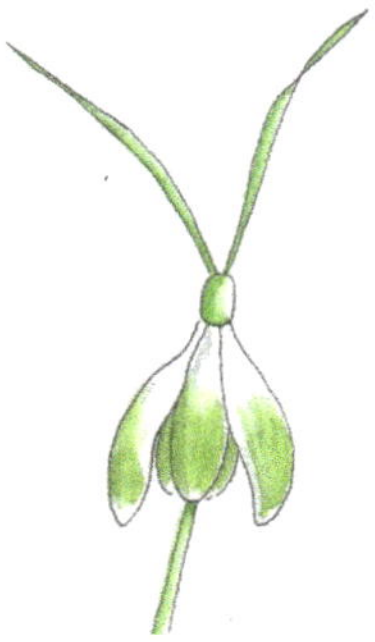

'Freddy King' or 'Freddy's King'
Smaller *G. nivalis* with delicately green-marked flowers and slender green ovary. Leaves applanate, erect, slender, blue-green. Outer segments tapering to pointed apex, small merging green lines above apex to one third of segment. Inner segments broad green inverted 'U'-shaped mark above sinus diffusing on basal edge. 2021. 14cm.

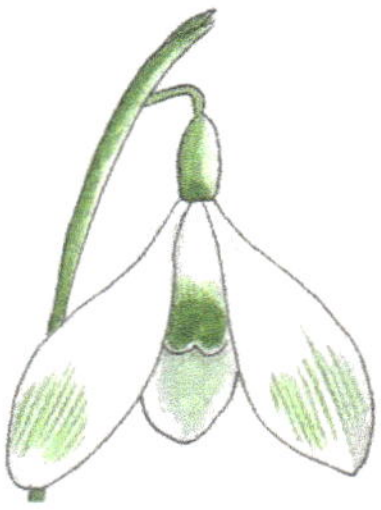

'Freddy's White Wonder'
Attractive, mainly white Poculiform flowers. Leaves erect, slender, grey-green. Segments roughly of equal length, rounded, tapering to bluntly pointed apex, occasionally small green marks above apex. Freddy Van Houtte. 15cm. 2018.

'Fridge Magnet'
Handsome, short *G. valentinei* with large, attractive flowers and good green mark suspended from long slender pedicel and slender, rounded triangular ovary. Leaves erect to splayed, grey-green. Outer segments clawed, tapering to pointed apex. Inner segments green inverted 'V'-shaped mark above sinus shading towards base. Vigorous and bulks up well. Andy Byfield. 2020. Winter/Spring. 8-10cm.

'Frohnauer Gold'
Beautiful yellow-marked *G. nivalis* with yellow pedicel and ovary, showing good colour from day one. Leaves applanate, erect, slender, blue-green. Outer segments slender, rounded to bluntly pointed apex. Inner segments, good yellow inverted 'V'-shaped mark above sinus running in lines towards base. Berlin. 2020. Winter/Spring. 15cm.

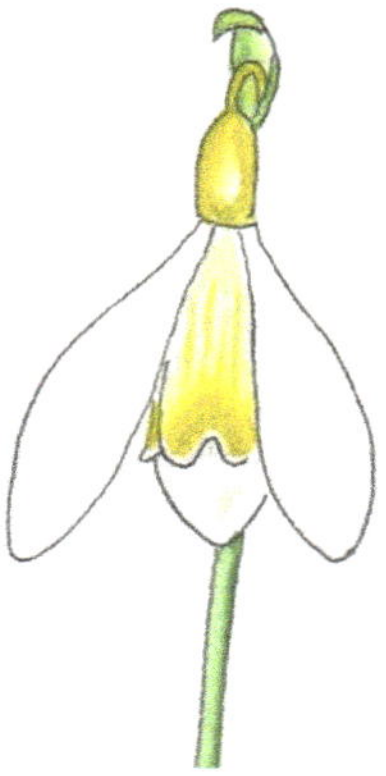

'Full Marks'
Quirky, rounded, green-tipped double *G. nivalis* clone with neat looking slender ovary. Leaves applanate, erect, slender, blue-green. Outer segments flattish with recurved margins tapering to bluntly pointed apex, small green marks above apex. Inner segments neat ruff with inverted mid-green 'V'-shaped mark above sinus. Some flowers produce three additional extra long segments similar to outers, extending from the centre of the ruff. Mark Brown and found in Shropshire.

'Fuzzy Wuzzy'
Attractive spiky snowdrop. Phil Cornish 2020.

'Gaia'
G. elwesii var. *monostictus* with slender looking, green-marked flowers and slender green ovary. Leaves supervolute, erect, broad, grey-green. Outer segments slender tapering to pointed apex, green-washed merging green lines above apex to three quarters of segment, shadowed on underside. Inner segments green mark above sinus almost to base, narrow white margin. Autumn/Winter. 2020.

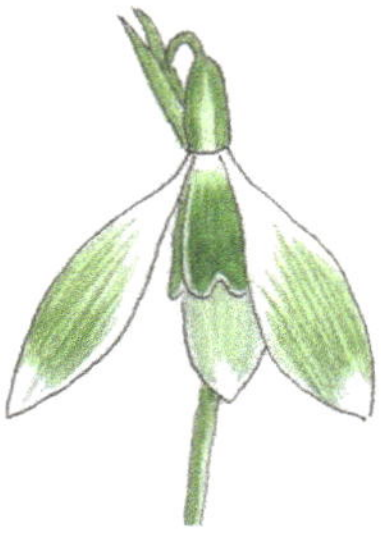

'Galloping Horses'
Good *G. elwesii* var. *monostictus* with erect scapes and elegant flowers. Leaves supervolute, broad, erect, recurved at tip, green. Outer segments tapering to bluntly pointed apex. Inner segments broad, good green heart-shaped mark above sinus. Winter/spring. 2022. 15-18cm.

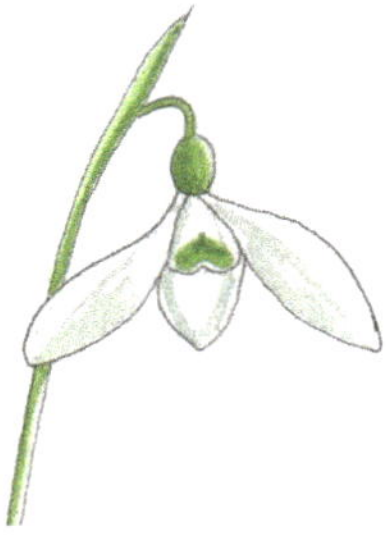

'Garden Peace'
Attractive flowers with light green spathe and ovary. Leaves broad, splayed, green. Outer segments rounded, tapering to pointed apex. Inner segments green mark above sinus merging into second mark towards base.

'Gekke Henkie'
Unusual looking flowers with flaring segments. Leaves medium, erect to arching, blue-green, paler median line. Outer segments flaring outwards, slender, upturned, narrow green mark above apex. Inner segments short, slender, green mark above sinus, orange anthers show through. Snowdropfevers. 2021. Winter/Spring. 15cm.

'Genainville'
Regular, double *G nivalis*. Leaves applanate, erect to arching, slender, blue-green. Outer segments flaring, slender, tapering to bluntly pointed apex. Inner segments neat ruff, narrow green inverted 'V'-shaped mark above sinus. Jean-Michel Groult and probably from Glenainville in Northern France. 2015. 15cm.

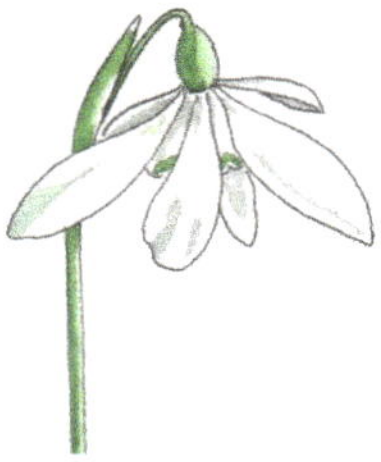

'Georgia Arula'
Neat, small *G plicatus* subs. *byzantinus* with attractive curling foliage. Leaves explicative, slender, blue-green, curled. Outer segments rounded, bluntly pointed at apex. Inner segments green mark above sinus. Strongly perfumed. Joe Sharman, Monksilver Nursery. Cambridge. Winter/Spring.

'Georgiana'
Well-proportioned *G. nivalis* with rounded flowers. Leaves applanate, erect, slender, blue-green. Outer segments rounded to bluntly pointed apex. Inner segments green mark above sinus. Richard Bashford.

'Gestrichelt Pulk'
Good example of a green-tipped snowdrop. Outer segments short claw, tapering to pointed apex, small green lines above apex. Inner segments green inverted 'V'-shaped mark above sinus. Bulks up well.

'Gimley Din'
Smaller flowers on erect scapes. Outer segments clawed, rounded to pointed apex. Inner segments small green mark above apex, no sinus.15cm.

'Gimley Durin'
Neat, well-rounded fully packed double flowers with good green marks, slender pedicel and slender, oblong olive-green ovary. Leaves erect, slender, blue-green. Outer segments flaring, rounded, incurved, tapering to bluntly pointed apex. Inner segments tight, neat ruff, broad, with inverted green 'V'-shaped mark above sinus. Paul Zazmierski and one of a series named after a character in Lord of the Rings.

'Glasdrum'
Leaves slender erect to splayed, blue-green. Outer segments slender rounded towards pointed apex. Inner segments narrow green 'U'- to 'V'-shaped mark above sinus, two paler green marks towards base. 12-15cm.

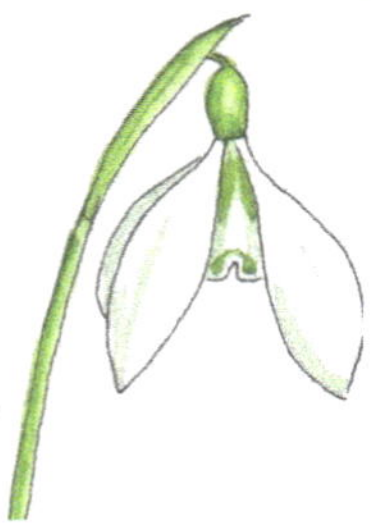

'Glauconite'
G. nivalis Virescent from *G.* 'Nova Gorica' x *G.* 'Green Tear'. Good green marked flowers with rounded green ovary. Leaves applanate, erect to arching, slender, blue-green. Outer segments rounded to bluntly pointed apex, green flush across segments with pronounced green lines above apex to base. Inner segments single mid-green mark above sinus diffusing at base, narrow white margin. A selection from Paul Barney, Edulis Nursery. Winter-Spring.

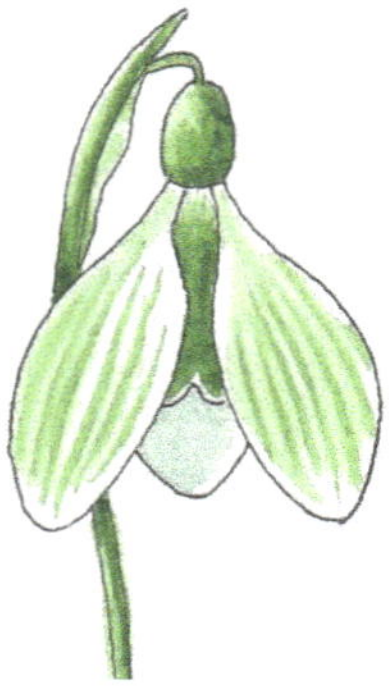

'Glen Una'
Shorter growing *G. elwesii* of Irish origin with short, erect scapes and large flowers. Leaves supervolute, broad, grey-green. Outer segments long, rounded, tapering to rounded apex, lightly longitudinally ridged. Inner segments broad green heart-shaped mark above sinus. Winter. 8cm.

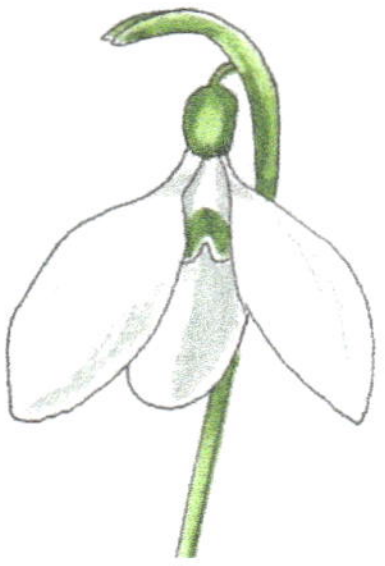

'Glooming'
Taller *G. nivalis* cultivar. Leaves applanate, erect to arching, slender, grey-green. Outer segments rounded, tapering to bluntly pointed apex, lightly longitudinally ridged, green lines running up ridges from above apex towards base. Inner segments green mark above sinus almost to base, narrow white margin. Gert-Jan van der Kolk, KAVB registered 16th January 2001. Winter. 18cm.

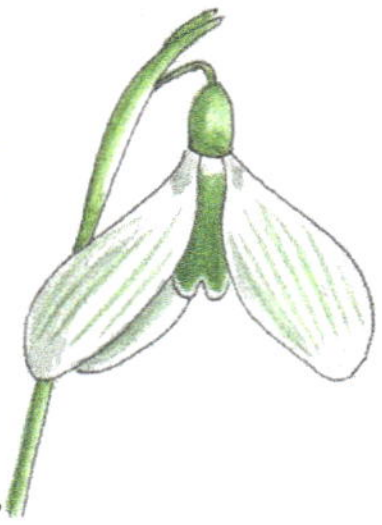

'Glory Gold'
Yellow marked flowers with good yellow pedicel and ovary. Leaves erect to splayed, slender, blue-green. Outer segments rounded to bluntly pointed apex. Inner segments good yellow mark. 10cm.

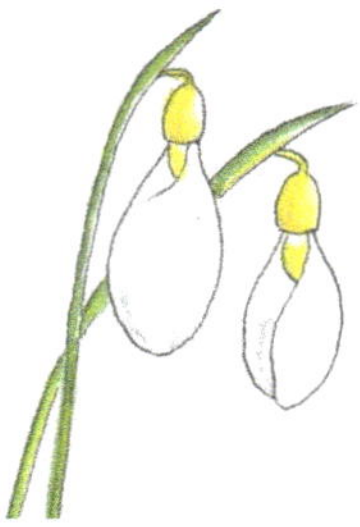

'Glowing'
Shapely slender flowers with oblong green ovary. Leaves erect to arching, slender, blue-green. Outer segments long, slender, tapering to pointed apex, longitudinally ridged. Inner segments long green inverted 'V'-shaped mark above sinus diffusing into yellow on basal edge. Winter/ Spring. 15cm.

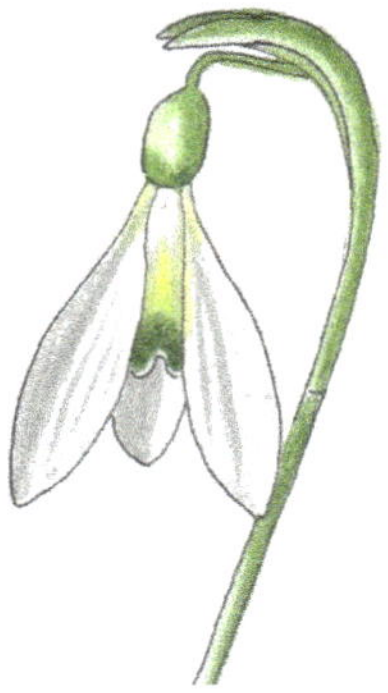

'Gold n Green'
Shorter growing but heavily yellow marked snowdrop on yellow scapes with yellow ovary and yellow-green spathe. Leaves medium, splayed, green. Outer segments rounded to bluntly pointed apex. Inner segments unusual yellow-green mark above sinus diffusing towards base. 8cm.

'Golden Acre'
Rare snowdrop with slender gold-tinged flowers on erect scapes, small, rounded yellow-green ovary and long, arching pedicel. Leaves slender, yellow-green. Outer segments slender, tapering to pointed apex. Inner segments pale yellow-green mark above sinus diffusing on basal side. Not a strong grower. From 'Snowdrop Acre', Norfolk. 2021. Winter/Spring. 15cm.

'Golden Eagle'
G. nivalis Sharlockii with long split spathe, long, slender, angled yellow pedicel and yellow ovary. Leaves applanate, erect, slender, green, paler median line. Outer segments rounded to bluntly pointed apex. Inner segments yellow mark above sinus. Originally *G.* 'Ecusson d'Or', after its first year it mutated into this variation and has been stable over a period of years. Brian Ellis. Winter/ Spring. 12cm.

'Golden Girl'
Long slender flowers with oblong, golden-green ovary and pedicel. Outer segments long, slender, rounded and tapering to pointed apex, lightly longitudinally ridged. Inner segments green Chinese bridge shaped mark above sinus paling into gold on basal edge. Cornelius Rosingh.

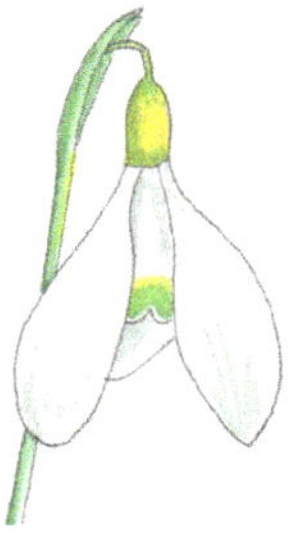

'Golden Goose'
Erect scapes with long rounded ovary. Leaves slender, erect. green. Outer segments flattened, rounded to apex, green mark above small sinus merging into second mark at base, yellow tinge. Inner segments green mark apex to base. 10cm.

'Golden Plummet'
Very pendulous flowers from long, slender, angled pedicel. Outer segments rounded to rounded apex, very lightly longitudinally ridged. Inner segments green mark above sinus diffusing into second mark towards base. Graham Gough. Winter/ Spring.

'Golden Tears'
Very rare, distinctive and expensive *G. plicatus* Inverse Poculiform snowdrop with erect scapes, yellow-green pedicels and rounded, yellow-green ovary. Leaves explicative, green-grey. Outer segments broad, rounded, reflexed along edges, lightly longitudinally ridged, broad yellow green mark above apex. Bred by Joe Sharman with the same parentage as *G.* 'Golden Fleece' but looks quite different. Very vigorous and can have twin flowers. 20-25cm. Sold on eBay in 2022 for £1850.

'Goldfinch'
Slender, green-marked flowers with oblong green ovary. Leaves erect, slender, green. Outer segments slender, tapering to pinched, pointed apex, light-green marks above apex on lower third of segment. Inner segments green inverted 'V'-shaped mark above sinus. Freddy Van Houtte. Winter/Spring. 15cm.

'Goldfinger'
Slender, yellow marked flowers with yellow-green scape and spathe, yellow pedicel and slender yellow ovary. Outer segments slender, rounded, tapering to pointed apex, Inner segments narrow inverted yellow 'V'-shaped mark above sinus.

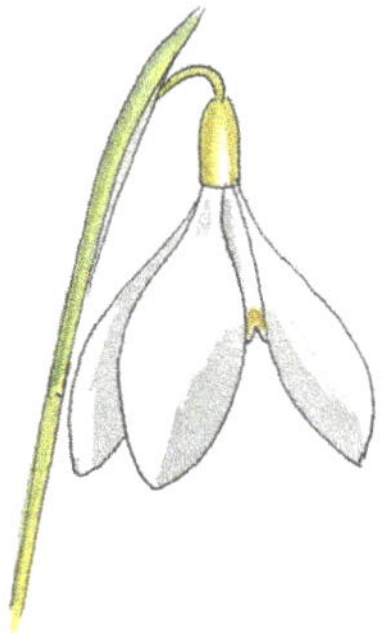

'Goldheart'
G. nivalis with erect scapes, slender pedicel and oblong green ovary. Leaves applanate, erect, slender, blue-green. Outer segments rounded, tapering to bluntly pointed apex. Inner segments narrow, light- green mark above sinus, distinctive golden colouring to base. Named by John Morley, North Green

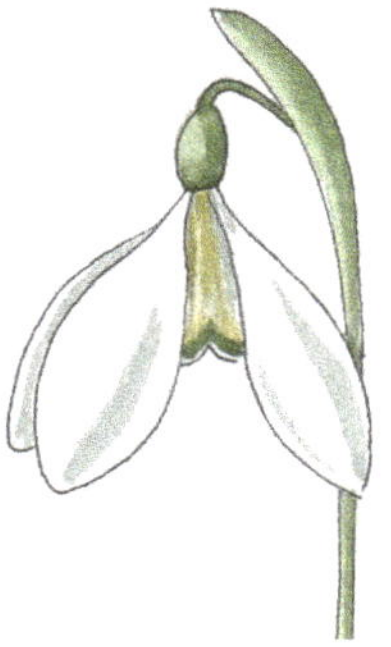

'Goldilocks'
Elongated flowers with short pedicel beneath curving spathe. Outer segments long, tapering to bluntly pointed apex. Inner segments yellow-green horseshoe shaped mark above sinus.

'Goldish'
G. platyphyllus with distinctive 'burnt orange' inner marking, lightly textured segments, rounded yellow-green ovary and slender flowers. Leaves supervolute, very broad, erect, green. Outer segments slender, tapering to pointed apex. Inner segments burnt orange mark above virtually no sinus, green mark at base running down lines towards apex. Colin Mason. Winter/Spring. 16-18cm.

'Goldsmith'
Elegant and delicate flowers on erect scapes with golden-yellow colouring and oblong yellow ovary, pedicel and spathe. Leaves erect, slender to arching, green. Outer segments long, slender, tapering to pointed apex, lightly longitudinally ridged. Inner segments slender, yellow inverted 'U'- to 'V'- shaped mark above sinus. Winter/Spring.

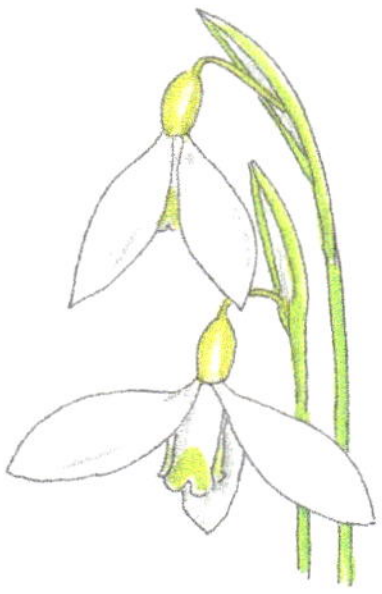

'Gore Booth'
Large flowers on shorter scapes with long triangular ovary. Leaves short at flowering, broad, erect, pointed, grey-green. Outer segments, clawed, pronounced shoulder, long, slender, tapering to pointed apex. Inner segments narrow, spreading inverted 'V'-shaped mark above small sinus. Sir Josslyn Gore Booth (1869-1944) of Lissadell Gardens on the shores of Sligo Bay, Ireland. The gardens originated in the 1760s, eventually abandoned and nature allowed to take over between1955 and 2003 when they were resurrected. Winter/Spring. 12cm.

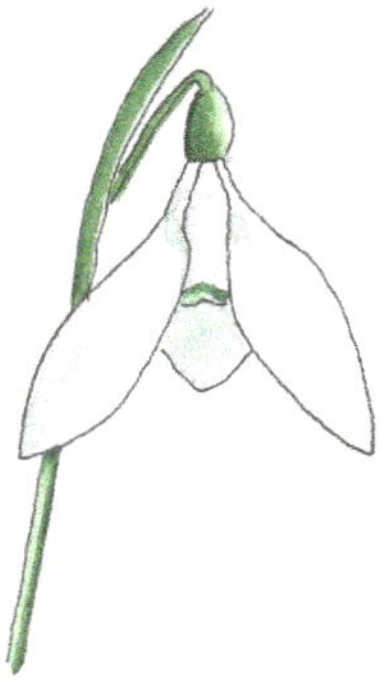

'Gorkains'
Rounded flowers with oblong, green ovary. Outer segments slender claw, broad, rounded to pointed apex. Inner segments green Chinese-bridge shaped mark above sinus.

'Goteborg'
G. sandersii group with large, long flowers on slender yellow pedicel and yellow-green spathe. Outer segments, long, incurved to rounded apex, lightly longitudinally ridged. Inner segments long with good golden-yellow inverted 'V'-shaped mark above sinus. Winter/Spring.16cm.

'Gothenburg Double'
G. Atkinsii Scharlockii double snowdrop with long split spathe. Leaves slender, erect, blue-green. Outer segments bluntly pointed at apex, green mark above apex. Inner segments rounded, neat ruff with inverted green 'V'-shaped mark above sinus. Winter/Spring.15cm.

'*gracilis* Kew Form'
Well-rounded flowers with oblong green ovary. Leaves applanate, erect, slender, green. Outer segments broad, rounded to bluntly pointed apex. Inner segments split green mark either side of sinus, second mark towards base.

'Grakes Oddity'
Unusual and quirky *G. plicatus*, every flower in the group different with very variable segments and markings. Leaves explicative, erect to splayed, slender, green, paler median line. Similar but smaller than *G.* 'Clovis' Very vigorous. From a seedling in *G.* 'Grake`s Heredij'. Valentin Wijnen, Grakes Garden, Belgium. 2019. Autumn/Winter.

'Green Bear'
Rounded, green marked, shorter *G. elwesii* Poculiform flowers. Leaves supervolute, erect to arching, broad, grey-green. All segments of equal length, broad, rounded to pointed apex, small green mark above apex. Cal Mateer, British Columbia.

'Green Besom'
Quirky snowdrop with long, slender, Scharlockii type split spathe and long, narrow, slender flowers. Leaves erect to splayed, slender, bright green. Variable slender, incurved segments with green marks above apex. 2020. European origin. Winter. 10-15cm.

'Green Bonnet'
Small delicately green-marked flowers with slender green ovary. Leaves slender, erect to arching, green. Outer segments broad, rounded, tapering to pointed apex. Light longitudinal ridging with green lines. Inner segments slender green inverted 'V'-shaped mark above sinus. Winter/Spring. 10cm.

'Green Dance'
Early snowdrop with delicate, spreading flowers and rounded green ovary. Outer segments slender, spreading, lightly longitudinally ridged, pale yellow-green wash on lower half of segments diffusing into lines. Inner segments dark-green broad roughly 'W'-shaped mark above sinus. Autumn.

'Green Day'
Virescent *G. nivalis* with slender looking flowers. Leaves applanate, slender, splayed, blue-green, paler median line. Outer segments slender, tapering to pointed apex. Green marks above apex to three quarters of segment. Inner segments solid mid-green mark from above sinus almost to base. Decora Nursery, Croatia. 12cm.

'Green Desire'
Long, slender, green-marked flowers with smaller rounded green ovary. Outer segments long, slender, incurved, rounded to pointed apex, pronounced green lines above apex to three quarters of segment. Inner segments mid-green mark above sinus almost to base. Richard Meijndert.

'Green Early Jewel'
G. elwesii var. *monostictus* with slender looking flowers. Leaves supervolute, medium, arching, ridged, grey-green. Outer segments slender tapering to pointed apex Inner segments single green mark above sinus almost to base, narrow white margin. Named by Oliver Vico and Ruben Billiet. 2021. Winter/Spring. 16cm.

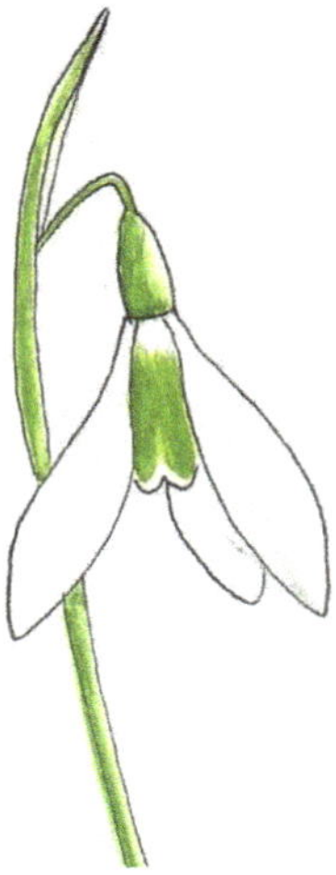

'Green Fairy'
Delicate-looking *G. nivalis* with green marked flowers. Leaves applanate, erect, blue-green. Outer segments long, rounded, bluntly pointed at apex, pronounced green stripes from above apex to base with delicate green wash. Inner segments darker green mark from above sinus to base. Said to have originated in Slovenia. January-February 10cm.

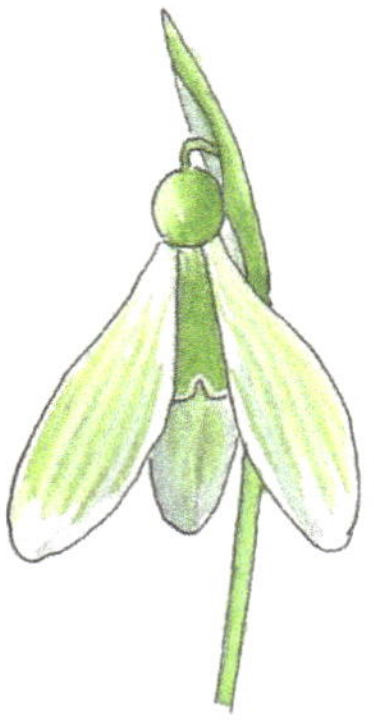

'Green Flight'
Attractive, *G. peshmenii* with lightly textured, green-tipped flowers suspended beneath slender, oblong green ovary. Outer segments rounded, lightly textured, tapering to pointed apex, small green mark above apex. Inner segments slender with emerald green 'V'-shaped mark above sinus.

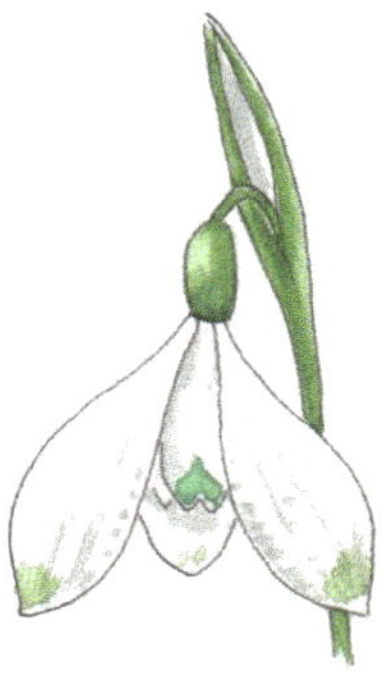

'Green Genes'
Elegant snowdrop with good green-marked flowers. Outer segments rounded to bluntly pointed apex, lightly longitudinally ridged, merging green lines from above apex to half of segment. Inner segments slender, mid-green heart-shaped mark above sinus.

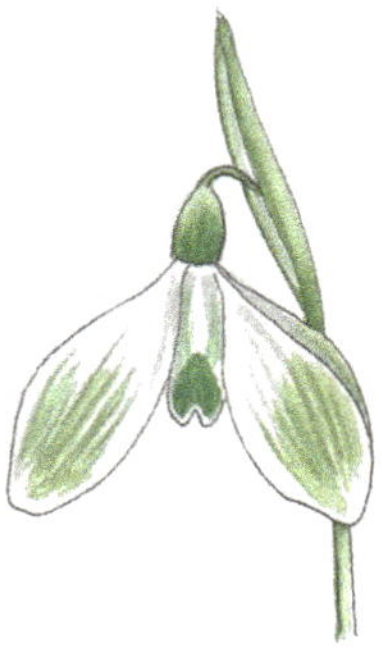

'Green Hell'
Small, heavily green-marked flowers with rounded green ovary. Outer segments incurved to pointed apex, heavily marked green above apex to base. Inner segments green mark across segment from above sinus to base. Uwe Stiebritz. 2021. Winter/Spring.

'Green Jewel'
An early flowering Virescent *G. nivalis* usually in flower before Christmas. Outer segments slender, tapering to pointed apex. Inner segments broad green mark across segment from above sinus almost to base, narrow white margin. Earlier and taller than *G.* 'Limetime'. Autumn-Winter.

'Green Lampshade'
Vigorous *G. plicatus* Inverse Poculiform snowdrop with flaring, heavily green marked flowers. Leaves explicative, green-grey. Outer segments broad, flattened, rounded at apex, green mark from above apex to base diffusing at basal edge. Inner segments green mark above sinus to base. Joe Sharman, 2021. Winter/Spring.

'Green Longfinger'
Distinctive long, slender flowers with incurved segments. Leaves erect, slender, green. Outer segments very long, slender, incurved, pointed apex with small green lines above apex. Inner segments broad dark-green inverted 'U'-shaped mark above sinus. 12cm. Winter/Spring.

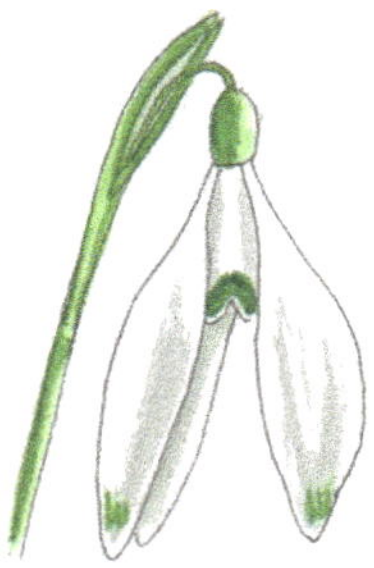

'Green Moon'
Small, neat *G. nivalis*. Leaves applanate, erect to splayed, slender, blue-green. Inner segments with crescent moon shaped mark above sinus. Ian Christie. 2018.

'Green Peacock'
Smaller snowdrop with heavily green marked flowers. Leaves erect to arching, slender, green. Segments roughly of equal length, slender, tapering to bluntly pointed apex, heavily marked green apex towards base. Winter/Spring. 10-12cm.

'Green Spring Fling'
G. nivalis with heavily green-marked flowers beneath long, oblong rounded green ovary. Leaves applanate, erect, slender, blue-green. Outer segments rounded, incurved to bluntly pointed apex, strong green lines above apex almost to base with green wash. Inner segments green mark above sinus to base, narrow white margin. Ruben Billiet. 2022. 16cm.

'Green Testament'
Charming, small, green-marked, double *G. nivalis* with long slender pedicel from between Sharlockii type split spathe and long, slender, green ovary. Leaves applanate, erect, slender, blue-green. Outer segments slender, incurved to bluntly pointed apex, green above apex diffusing towards base. Inner segments, slightly shorter than outer, incurved to bluntly pointed apex and similarly green marked to outer. Freddy Van Houtte. Winter/Spring. 10cm.

'Green Tiger'
Slender-looking, well-proportioned flowers with good green colouring. Outer segments rounded tapering to pointed apex, green lines above apex almost to base, light-green wash. Inner segments broad green mark above sinus, fading across segment almost to base, two darker ovals towards base, narrow white margin. Winter/Spring. 16cm.

'Green-Tipped Balloon'
Shorter growing snowdrop with large, well rounded, balloon-like flowers. Leaves medium, erect to arching, grey-green. Outer segments rounded tapering to bluntly pointed apex, lightly textured, small green marks above apex. Inner segments, green mark across segment from above sinus almost to base. 10-12cm.

'Green-Tipped Orange Star'

Quirky-looking later-flowering snowdrop suspended from angled pedicel and small triangular green ovary. Outer segments very long, slender, bluntly pointed at apex, small green mark above apex. Inner segments variable with small green mark, clearly revealing orange anthers. Ruben Billiet. 2022. February/March.

'Green Tornadoes'

Unusual looking, small, double flowers with split spathe, narrow green ovary and additional, extra-long, protruding segments. Leaves erect to splayed, slender, blue-green, paler median line. Outer segments very green, slightly ruffled at edges with longer segments protruding and reflexing. Inner segments green mark above sinus to base. Freddy Van Houtte. 2019. Winter/Spring. 12cm.

'Green Trym'

Another selection from The Patch, the late Margaret Owen's garden in Shropshire. Chosen and named for its very green, broad arching leaves. Small rounded ovary and good *G.* 'Trym' type flowers. Outer segments rounded lightly longitudinally ridged, good green heart-shaped mark above sinus. Sadly very slow to bulk up.

'Green Whisp'

Slender, heavily green-marked flowers that barely open. Leaves erect, slender, blue-green. Outer and inner segments similar, slender, tapering to bluntly pointed apex, strong green mark above apex diffusing and shading green towards base. A whisp is a collective noun for a flock of snipe. Originally from Johan Mens who passed it to the late Veronica Cross and from thence to John Morley, North Green Snowdrops who bulked up the bulbs. Winter/Spring.10cm.

'Greengage'

This *G. plicatus* Inverse Poculiform stands out with its bold green marking and attractive green foliage. Leaves explicative, green. Outer segments flattened, rounded at apex, lightly longitudinally ridged, light-green mark above apex. Inner segments broad green mark above sinus to half of segment. Olive Mason, from her garden. Winter.

'Greenpeace'

Tall, well-marked *G plicatus* cultivar with large, well-rounded, textured flowers, regularly producing two scapes to each bulb. Leaves explicative, erect to arching, green-glaucous, paler median line. Outer segments clawed, broad, rounded to bluntly pointed apex, goffered at base, lightly longitudinally ridged and textured. Inner segments broad, straight-sided, mid-green mark above sinus almost to base, narrow white margin. Good perfume. Derek Fox from Oliver Wyatt's former garden, Naughton, Suffolk. January/February. 18cm.

'Grizzly'

G. gracilis. Syn *G. graecus*.
Well-shaped flowers with oblong green ovary. Leaves applanate, medium, slender, green. Outer segments light goffering on basal edge, tapering to pointed apex, green lines and light-green wash above apex to base. Inner segments mid-green mark across segment from above sinus almost to base, narrow white margin. 12cm.

'Grössenhof'

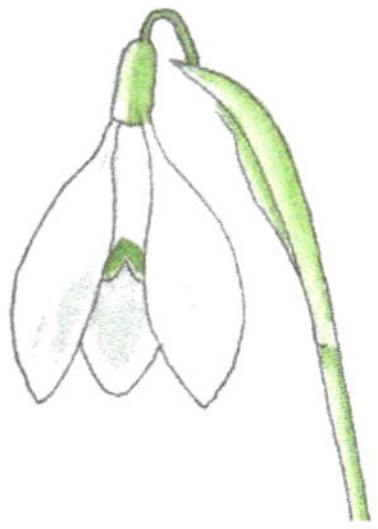

Slender, rounded flowers on erect, slender pedicel with slender, oblong green ovary. Outer segments, clawed, rounded, tapering to pointed apex. Inner segments, narrow green inverted 'V'-shaped mark above sinus.

'Grumpy's Brother'

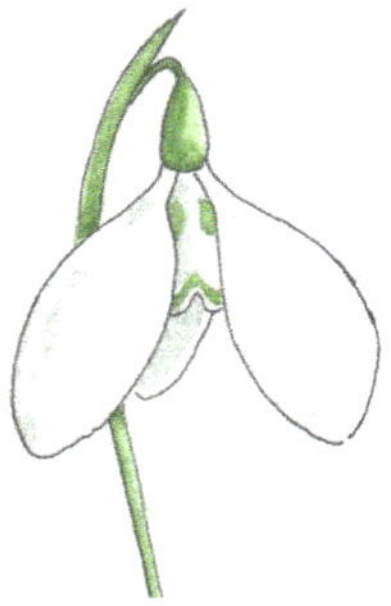

Similar looking snowdrop to *G.* 'Grumpy' but this one is a true *G. plicatus* having slender ovary and long slender flowers. Leaves explicative, grey-green. Outer segments long, tapering to bluntly pointed apex. Inner segments typical *G.* 'Grumpy' mark with characteristic downturned mouth and two eye marks towards base. Patricia Elkington, Hampshire.

'Grüne Donauzwerg'

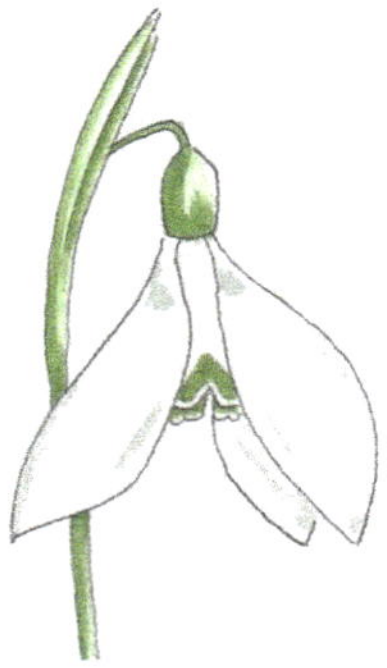

Slender, rounded flowers with oblong green ovary. Outer segments rounded, tapering to pointed apex. Inner segments inverted green 'V'-shaped mark above sinus.

'Grüne Heike'

Unusually shaped, green marked flowers on erect scapes with slightly inflated spathe and rounded green ovary. Outer segments, flat, broad, clawed, slender, incurved, flaring upwards, rounded to bluntly pointed apex, narrow green line running up centre of segment from above apex to two thirds of segment. Inner segments short, revealing anthers, flushed orange, green mark above sinus. Winter/Spring. 12-15cm.

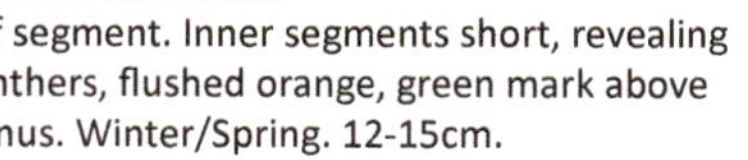

'Grüne Nase'
Erect spathe and long arching pedicel. Outer segments slender, rounded tapering to bluntly pointed apex. Inner segments long to two thirds of outer, narrow, green dot either side of small sinus. Winter/spring. Gerard Raschun. 2022.

'Grüne Spitze'
Early *G. elwesii* with well rounded and textured flowers and small rounded green ovary, usually flowering before Christmas. Leaves supervolute, broad, erect to arching, matt grey-green. Outer segments broad, rounded, tapering to bluntly pointed apex, lightly longitudinally ridged and textured, short green lines above apex. Inner segments green mark above sinus. Staudengartnerei Jentsch. November/ December. 12cm.

'Grünspecht No 4'
Large, rounded flowers suspended from long slender pedicel and oblong, rounded, green ovary. Outer segments broad, rounded, tapering to pointed apex, longitudinally ridged, light-green marks above apex. Inner segments broad green 'U'-shaped mark above sinus.

'GRY2'
Small, very slender flowers with slender, oblong green ovary. Outer segments very slender, tapering to pointed apex. Inner segments small green mark above sinus.

'Gwawis Caribou'
Small *G. elwesii* Poculiform snowdrop with rounded flowers of roughly equal segments and small rounded green ovary. Outer segments rounded, rounded at apex, lightly longitudinally ridged, small green mark either side of apex. Inner segments almost as long as outer, green mark above apex. Cal Mateer, British Columbia, 2022. Winter.

'Haconby Early'
Good, early flowering *G elwesii* var. *monostictus* with slender, well shaped flowers. Leaves supervolute, erect to arching, broad, grey-green. Outer segments slender, slender claw, tapering to bluntly pointed apex, inner segments, small heart-shaped green mark above sinus. Cliff Curtis, Haconby, Lincolnshire. December/January. 15cm.

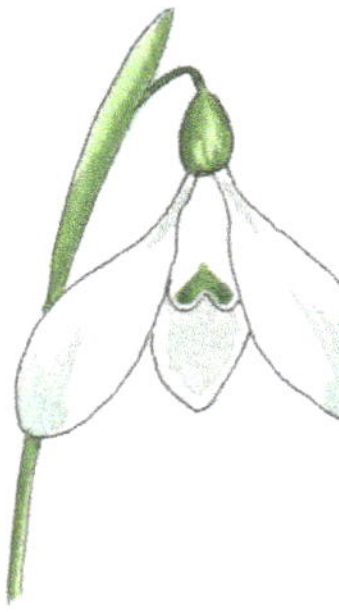

'Halcyon'
Rounded flowers with oblong green ovary. Outer segments broad, rounded to rounded apex. Inner segments green Chinese-bridge shaped mark above sinus.

'Hanging Out'
This snowdrop really does 'hang out' from between its Scharlockii-type split spathe with flowers suspended from long, slender, arching pedicel. Leaves erect to arching, slender, green. Outer segments rounded to bluntly pointed apex, green shading above apex to half of segment. Inner segments green inverted 'V'-shaped mark above sinus. Valentin Wijnen, Grakes Heredij. Winter/Spring. 2019. 16cm.

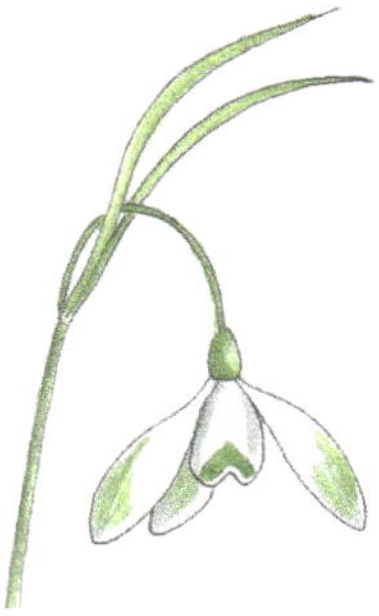

'Hardwick'
Small-flowered *G.plicatus* hybrid cultivar with delicate looking, rounded flowers on erect scapes. Leaves explicative, grey-green, paler median line. Outer segments rounded tapering to bluntly pointed apex. Inner segments green inverted 'V'-shaped mark above sinus. Glad Patterson, Tasmania.

'Harley Milne'
Large, substantial-looking, very white, cup-shaped, *G. plicatus* subsp. *byzantinus*. Leaves plicate, broad, shiny green. Outer segments very broad, incurved and rounded at apex. Inner segments lightly longitudinally ridged with well shaped mid-green mark above sinus and second mark towards base. Named in memory of a past president of the SRGC. A selection originally from the Brechin Castle snowdrops by Ian Christie. 2018. 10-15cm.

'Haroula'
Small *G. reginae-olgae* subs. *reginae-olgae*. Attractve smaller Autumn flowering snowdrop. Leaves applanate, erect, slender, grey-green, paler median line, small or absent at flowering. Outer segments slender, rounded, bluntly pointed at apex. Inner segments, green inverted 'V'-shaped mark above sinus. Melvyn Jope. September/October. 12cm.

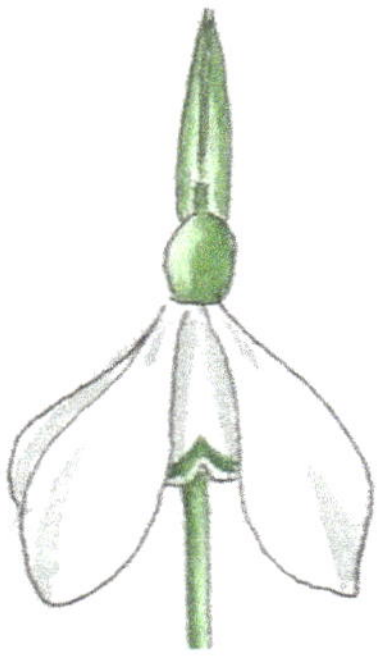

'Harpur Grace'
G. bursanus with erect spathe, long, slender yellow-green ovary and scape. Leaves applanate-narrowly explicative, blue-green, short at flowering. Outer segments short claw, incurved, rounded towards apex. Inner segments yellow-green heart-shaped mark above sinus merging into second oval towards base. Autumn. 10cm.

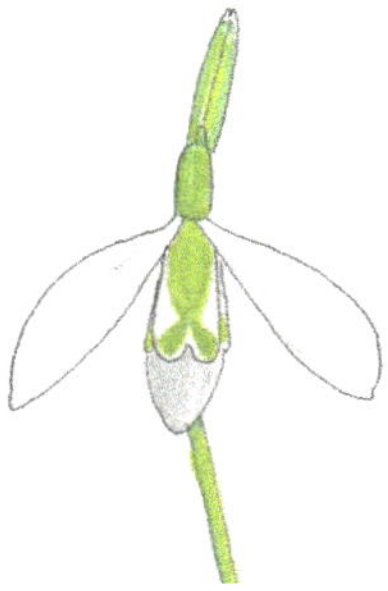

'Hattrick'
G. nivalis Sharlockii snowdrop with split spathe, erect pedicel, and slender, heavily green-marked flowers. Leaves applanate, erect, slender, grey-green. Outer segments very slender, bluntly pointed at apex, mid-green mark across segment from apex to base. Inner segments, long green mark above sinus diffusing towards base. Freddy Van Houtte. Winter/Spring. 15cm.

'Hawk'
Green marked *G. elwesii* with rounded flowers and slender green ovary. Leaves supervolute, broad, erect, hooded, grey-green, Outer segments rounded, tapering to pointed apex, small merging green lines above apex on lower quarter of segment. Inner segments green inverted 'V'-to 'W'- shaped mark above sinus and second oval towards base. Richard Bashford. 2020. Winter/Spring. 16cm.

'Heaven's Patcch'
Very white looking, rounded flowers shown off to perfection against attractive glossy bright-green foliage. Leaves medium, splayed, bright glossy green. Outer segments rounded to bluntly pointed apex. Inner segments small light-green mark above sinus. When the flower is closed it gives the appearance of being an albino snowdrop. Winter/Spring. 10cm.

'Heloise des Essourts'
G. nivalis double with smaller, rounded, green-marked flowers and small, triangular green ovary. Leaves applanate, erect to arching, slender, green-blue. Outer segments rounded, tapering to bluntly pointed apex, small green marks above apex. Inner segments spreading inverted green 'V'-shaped mark above sinus. Jean-Luc Panier, St Germain des Essourts, Normandy, and named for his youngest daughter. 2009.

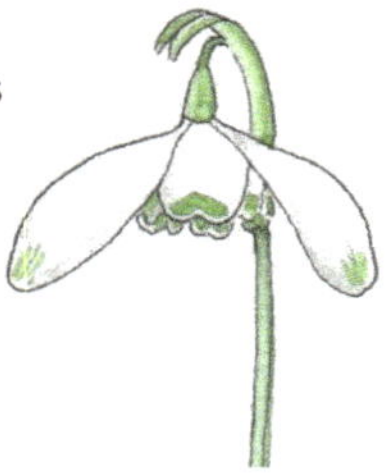

'Henley Deep Green'
Shorter-growing *G. elwesii* with good green inner mark. Leaves supervolute, broad, erect to arching, grey-green. Outer segments rounded to bluntly pointed apex. Inner segments green mark above sinus diffusing slightly at base, narrow white margin. Similar in appearance to *G.*'Brian Matthew'. Paul Barney, Edulis Nursery. 2012.

'Herzilein'
Smaller snowdrop with long slender, yellow-green ovary. Leaves slender arching blue-green. Outer segments, slender, short claw, prominent shoulder, flaring upwards, pointed at apex. Inner segments broad, rounded with good heart-shaped green mark above sinus. Germany. Autumn. 10-15cm.

'Hidden Secret'
An unusual but standard-looking *G. nivalis* double with a second row of inner segments 'hidden' beneath the first. Leaves applanate, erect to splayed, slender, blue-green. Outer segments tapering to bluntly pointed apex. Inner segments green mark above sinus, second row of segments with similar marks 'hidden' beneath the first, difficult to see unless the flower is turned upwards, hence the name. Mark Brown. 15cm.

'High Tide'
Large triangular green ovary. Outer segments flattened, rounded to apex, broad green mark above apex to half of segment. Inner segments narrow, broad green mark above sinus.
Winter/Spring.

'Hillview'
G. plicatus with long, rounded flowers suspended from slender pedicel with small, oblong green ovary. Leaves plicate, erect, green-grey. Outer segments rounded tapering to bluntly pointed apex. Inner segments green Chinese-bridge shaped mark above sinus. Winter/Spring.

'Honeypie'
Slender, rounded green-marked flowers with slender green ovary. Leaves, slender, splayed, blue-green. Outer segments, clawed, rounded, tapering to pointed apex, yellow-green mark above apex. Inner segments good green heart-shaped mark above sinus merging into second waisted mark to base. Small bulbs and plants short in stature but very floriferous and good perfume. 2022. 8cm.

'Howick Starlight'
A slender Hybrid cultivar with delicate looking flowers and small rounded, triangular green ovary. Leaves erect, slender, green-blue. Outer segments slender, incurved to bluntly pointed apex, small merging yellow lines above apex. Inner segments broad with yellow-green inverted 'V'-shaped mark above sinus. Northumberland. January/February. 14cm.

'Ice Princess'
Said to be the first-ever double Poculiform snowdrop with huge, broad, rounded, mainly white double flowers, large fat buds and mainly ice-white segments. Outer segments white, flaring, well rounded. Inner segments slightly shorter than outer with variable green marks from year to year. Flowers seem to continue expanding as they grow. Short, vigorous and free-flowering. Joe Sharman, Monksilver Nursery, Cambridge. 2020. Winter/Spring. 8cm.

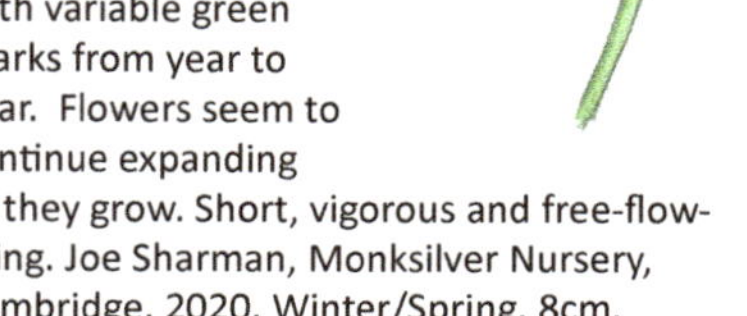

Italian Yellow'
Large, slender, yellow marked flowers with yellow tinged spathe, yellow pedicel and oblong yellow ovary. Leaves erect to arching, grey-green. Outer segments slender, tapering to pointed apex. Inner segments yellow inverted 'V'-shaped mark above sinus. Italy. Peter Erskine. Winter/Spring. 15cm.

'Jack By The Hedge'
Elegant looking *G. elwesii* with good green marks. Leaves supervolute, erect, broad, grey-green. Outer segments rounded to pointed apex, strong green marks above apex. Inner segments green inverted 'V'-shaped mark above sinus bleeding on basal edge. Paul Barney, Edulis Nursery. Ashampstead. Vigorous. 2011. 30cm.

'Janet Cropley'
A dainty, small-flowered, green-tipped *G. nivalis*. Leaves applanate, erect to arching, slender, blue-green. Outer segments tapering to bluntly pointed apex, unusual green inverted 'U'-shaped mark above apex. Inner segments, small green inverted 'U'- to 'V'- shaped mark above sinus. Richard Bashford and Valerie Bexley, Woodchippings, Northamptonshire

'Jasper'
G. nivalis with long, slender flowers. Leaves applanate, erect, slender, blue-green. Outer segments paddle-shaped, very long, slender, tapering to bluntly pointed apex, reflexing upwards at edges, merging light-green stripes above apex towards base. Inner segments broad inverted green 'U'-shaped mark above sinus. Snowdropfevers. Winter/Spring. 15cm.

'Je Suis Charlie'
Delicate-looking flower with rounded green ovary. Outer segments slender, rounded to bluntly pointed apex. Inner segments rounded green mark above sinus.

'Joe's Poculiform'
Slender-looking, white Poculiform flowers. Segments of equal length. slender, rounded, tapering to pointed apex. Joe Sharman, Monksilver Nursery, Cambridge.

'John Morley'
Smaller Galanthus hybrid with very large, handsome flowers on erect scapes having large mid-green ovary. Leaves erect to arching, slender, green-grey, paler median line. Outer segments broad, clawed, rounded to pointed apex, lightly longitudinally ridged. Inner segments darker green mark either side of sinus paling on basal edge and lightly joining paler, longitudinally divided mark towards base, can also have yellow shading. Originally thought to be *G. plicatus* subs. *byzantinus* or *G. gracilis* and first noticed at North Green as being extremely floriferous. 2022.

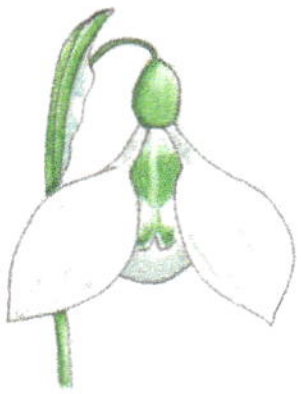

'John Owen'
Well-rounded flowers suspended from slender pedicel. Outer segments broad, rounded, tapering to bluntly pointed apex. Inner segments broad green mark above sinus merging into reversed mark above forming a roughly 'X'-shaped mark. From the late Margaret Owens garden, Acton Pigot, Shropshire and named for her son. Good grower and bulks up well.

'Judith'
Early *G. elwesii* var. *monostictus*. Leaves supervolute, broad grey-green, short at flowering. Outer segments rounded, short claw, tapering to pointed apex. Inner segments single green mark from above sinus almost to base with deep 'V'-shaped cut on basal edge. Autumn/Winter. 10cm.

'Julian Scholz'
G. nivalis with good green mark. Leaves applanate, erect, blue-green. Outer segments slender, rounded tapering to bluntly pointed apex. Inner segments mid-green mark across segment from above apex to base, narrow white margin. Named by Novy Bar, Northern Czech Republic, for Julian Scholz.

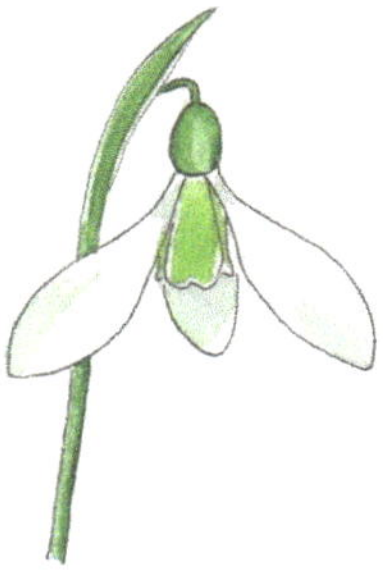

'Just White'
G. nivalis with very long, slender, pointed flowers. Leaves applanate, erect to arching, blue-green. Outer segments long, slender, tapering to pointed apex. Inner segments narrow green inverted 'U'-shaped mark above sinus.

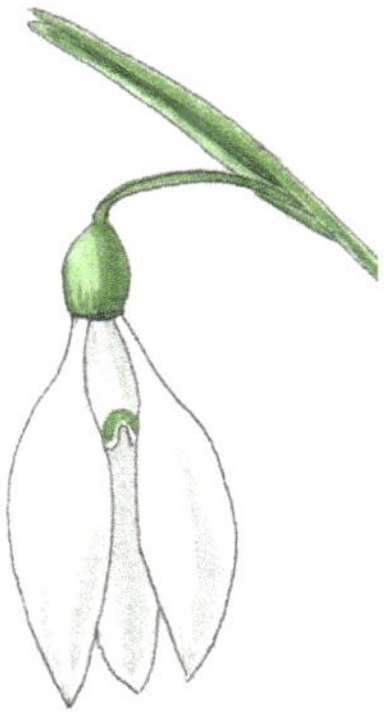

'Kaina'
Delicate-looking flowers suspended from slender pedicel, green-yellow spathe and pale yellow ovary. Outer segments slender, tapering to bluntly pointed apex. Inner segments inverted pale yellow 'V'-shaped mark above sinus.

'Kanari'
Shorter-growing *G. nivalis* with green scape and spathe, yellow pedicel and ovary. Leaves applanate, erect, slender, green-blue. Outer segments rounded tapering to pointed apex. Inner segments yellow-green 'W'-shaped mark above sinus. Autumn/Winter. 2021. 8cm.

'Katherine James'
Shapely buds opening into rounded flowers suspended from slender pedicel. Outer segments clawed, rounded to rounded apex. Inner segments green mark above sinus almost to base, narrow white margin.

'Kato'
Rounded green-marked flowers on slender pedicel with rounded green ovary. Outer segments rounded to bluntly pointed apex, short dark-green mark above apex. Inner segments dark-green inverted 'V'-shaped mark above sinus.

'Karst Dumpy'
Small stocky snowdrop originating in Italy. Edulis Nursery. 2020.

'Kaur'
Full, fine double in the Estonian Bird series of snowdrops. (Kaur = Arctic Loon - *Gavia arctica*). Erect scapes and slender spathe with small rounded triangular ovary. Outer segments full, slender claw, rounded, bluntly pointed at apex, usually unmarked. Inner segments variable, some white, some broad green mark above sinus and sometimes second pale mark at base. Often three additional segments between outers and inners which make the flower look like a double Poculiform. Vigorous and bulks up well. Tavii Tuulik, Estonia. 13cm.

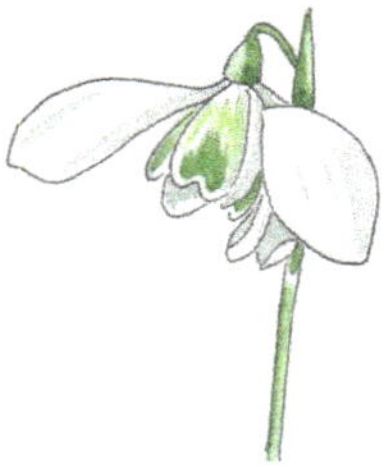

'Kelpie'
Yellow green pedicel and ovary. Outer segments rounded to pointed apex. Inner segments narrow green inverted Chinese-bridge shaped mark above sinus. A Kelpie is a mythical Scottish water horse which inhabits rivers and streams.

'Kencot Pickles'
Wispy-looking, green-marked double snowdrop. Outer segments, five-six, very slender, tapering to pointed apex, light-green mark above apex. Inner segments slender with pale-green inverted 'U'-shaped mark above small sinus.

'Kera'
A new *G. plicatus* with heavily textured, rounded flowers on erect scapes with rounded, slightly ridged green ovary. Leaves explicative, erect to arching, slender, green, short at flowering. Outer segments rounded, heavily longitudinally ridged and textured. Inner segments green mark above sinus. Taavi Tuulik, from the cultivar *G.* 'Suur Tilk', originally from Sulev Savisaar.

'Kermode Bear'
Good *G. elwesii* Poculiform flowers with even, regular segments. Leaves supervolute, erect to arching, broad, grey-green. Segments all roughly of equal length, slender, tapering to bluntly pointed apex, variable small green marks above apex. Cal Mateer, British Columbia.

'Ketskhovelii'
Lightly textured flowers beneath slender oblong ovary. Leaves broad, green, central median line. Outer segments lightly longitudinally ridged and textured, goffering at base, tapering to pointed apex. Inner segments. mid-green, squared, inverted 'U'- shape above sinus.

'Kill Abbey'
Erect scapes with slender spathe. Leaves erect to arching, ridged, green. Outer segments rounded to bluntly pointed apex, lightly longitudinally ridged and textured. Inner segments, dark-green heart-shaped mark above sinus. Winter/Spring. 16cm.

'Kiur'
Estonian 'Bird' Group of snowdrops 'Kiur - Tree Pipit - *Anthus trivalis*. Exciting new introduction of superb double snowdrops. Outer segments well rounded, tapering to rounded apex. Inner segments broad tight cluster with inverted green 'V'-shaped mark above sinus merging into broad mark to base. Introduced and bred by Tavii Tuulik, Estonia.

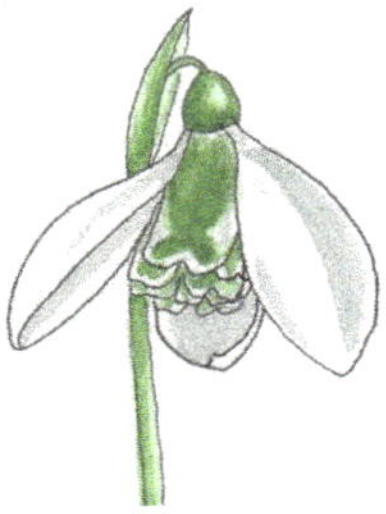

'Koksilah Caribou'
Small, neat *G. elwesii* Inverse Poculiform snowdrop with good green marking on all segments. Leaves supervolute, broad, erect to arching, grey-green. Segments longitudinally ridged, reflexed edges, four green marks on each segment, two above apex and two towards base.
Winter/Spring.

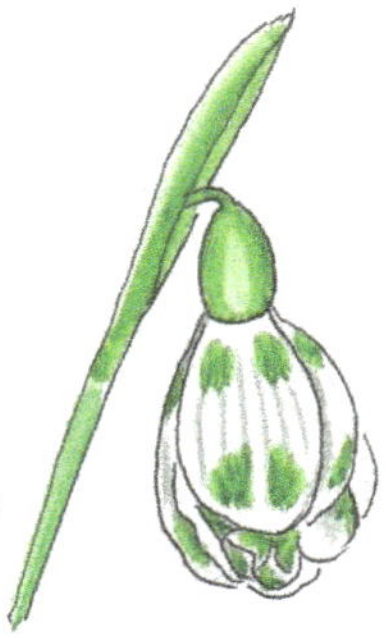

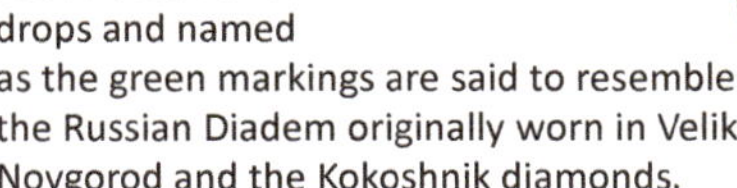

'Kokoshnik Diadem'
Large-flowered recent introduction making an imposing plant, although slow to mature. Rounded green ovary. Outer segments flattened, rounded at apex with good green mark above apex and second mark towards base. Inner segments dark-green mark across segment from above sinus, diffusing at base, narrow white margin. North Green Snowdrops and named as the green markings are said to resemble the Russian Diadem originally worn in Veliki Novgorod and the Kokoshnik diamonds.

'Komma'
Elegant-looking flowers suspended from green ovary, pedicel and spathe. Leaves erect, blue-green. Outer segments long, slender, incurved, slender claw, tapering to pointed apex. Inner segments broad, green mark above sinus. 15cm.

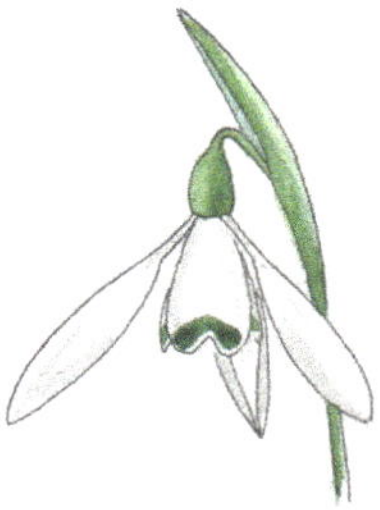

'Krim Wild' - ex 'Ann Repnow'
Tall erect scapes with long arching pedicel and well-shaped, textured flowers. Leaves broad, arching, green. Outer segments lightly longitudinally ridged and textured, tapering to bluntly pointed apex. Inner segments broad inverted green 'U'-shaped mark above sinus. Winter/Spring. 16cm.

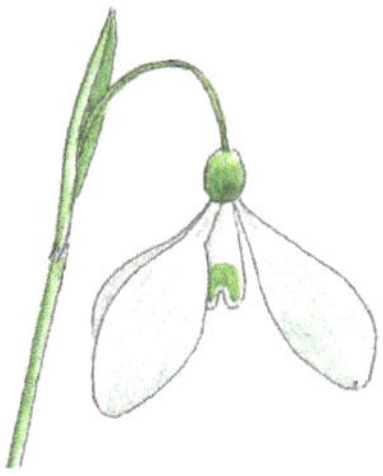

'Kudrus'
A small, attractive *G. nivalis* Estonian Spirit Group snowdrop with pale, yellow-green pedicel and ovary and rounded flowers. Leaves applanate, erect, slender, silver-grey. Outer segments rounded to bluntly pointed apex, short claw and pronounced shoulder. Inner segments, slender green inverted 'V'-shaped mark above sinus merging into paler yellow streak towards base. The Estonian Spirit series of snowdrops were discovered in abandoned gardens and woodland in Estonia and named for their pale markings. This name means tiny pearl. Taavi Tuulik, Estonia. 2022. 10cm.

'La Mornier'
Tightly packed, neat double snowdrop with long slender green ovary. Outer segments slender, flaring, tapering to pointed apex. Inner segments broad neat ruff, narrow green Chinese-bridge shaped mark above sinus.

'La Vie En Rose'
Attractive heavily green-marked flowers on slender pedicel with broad spathe and rounded, oblong green ovary. Outer segments broad, rounded, tapering to pointed apex, green lines above apex to three quarters of segment. Inner segments broad green mark above sinus and two paler ovals towards base.

'Lady Macbeth'
An USA selected *G. nivalis*. Tall, stately plants with large, weighty looking, well-proportioned flowers. Leaves applanate, medium, erect to arching, grey-green. Outer segments broad, rounded, rounded at apex. Inner segments dark-green inverted 'U'-shaped mark above sinus. Edgwood Gardens, USA. 2022. 25cm.

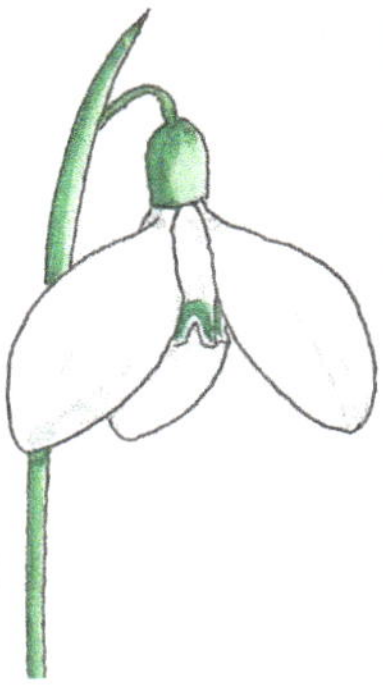

'Ladyboy'
Shorter growing *G. elwesii* snowdrop with large, rounded flowers on short erect scapes. Leaves supervolute, broad, short, silvery grey-green. Outer segments large, rounded to bluntly pointed apex, small green lines above apex. Inner segments broad dark-green inverted 'V'-shaped mark above sinus merging into second mark towards base. Andy Byfield. 10cm.

'Lagle'
Estonian bird series. (Lagle = Barnacle Goose - *Branta leucopsis*). Attractive, neat, well-rounded regular double flowers with slender triangular ovary. Outer segments broad, rounded, rounded at apex. Inner segments inverted green 'V'-shaped mark above sinus often merging into broader mark above, diffusing towards base. Tavii Tuulik, Estonia, 2020. 16cm.

'Lamplighter'
Creamy coloured flowers with oblong yellow-green ovary and long arching pedicel. Outer segments flushed cream, rounded to pointed apex. Inner segments narrow yellow inverted 'V'-shaped mark above sinus. *G. rizahensis* hybrid possibly crossed with *G. nivalis*. Jorg Lebsa.

'Lanky'
As its name implies, very elongated flowers with long slender ovary. Outer segments, slender, rounded, rounded at apex with green lines above apex to half of segment. Inner segments, small green heart-shaped mark above sinus. 2021.

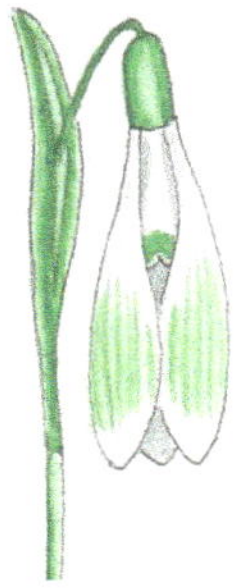

'Leilani'
Smaller *G.nivalis* with slender looking flowers, good green marks and small green rounded, triangular ovary. Leaves applanate, erect to splayed, slender, blue-green. Outer segments slender, tapering to pointed apex, green mark above apex. Inner segments slender, broad green inverted 'U'-shaped mark above sinus. Angelina Petrisevac. Winter/Spring. 8cm.

'Lemon'
G. elwesii with slender-looking flowers and rounded olive green ovary. Leaves supervolute, erect. broad, grey-green. Outer segments slender, tapering to bluntly pointed apex. Inner segments narrow, olive green mark above sinus.

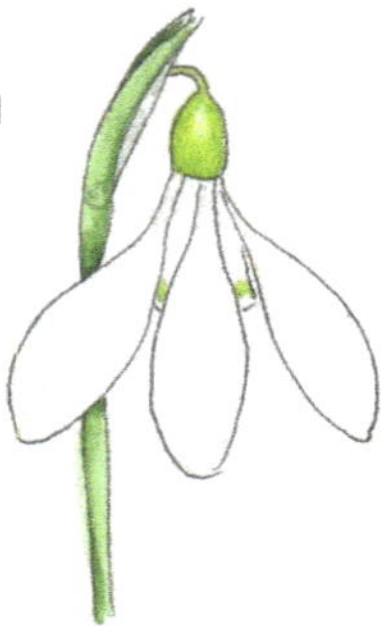

'Lemon and Lime'
Very attractive new contrasting yellow and green snowdrop with good strong colouring, green spathe, yellow-green pedicel and yellow ovary. Leaves erect, slender, blue-green. Outer segments rounded to bluntly pointed apex, lightly longitudinally ridged. Inner segments lime-green mark above sinus diffusing on basal edge.

'Lemon Kindness'
Shorter growing snowdrop with slender looking yellow, marked flowers, yellow ovary and pedicel. Leaves short at flowering, slender, erect, green. Outer segments slender tapering to pointed apex. Inner segments yellow inverted 'V'-shaped mark above sinus, diffusing on basal side. Igor Forester. 2020. 10cm.

'Liebsten'
Well-rounded, green marked flowers suspended from slender arching pedicel with oblong green ovary. Outer segments broad, rounded, rounded at apex, green inverted 'V'-shaped mark above apex. Inner segments very narrow green mark above sinus. Hans-Joachim Tillman.

'Lil'wat Caribou'
Small *G. elwesii* Inverse Poculiform snowdrop. Leaves supervolute, broad, incurved, pointed at apex, grey-green. Segments all of equal length, lightly longitudinally ridged, four green marks on segments, two joined above small sinus, two ovals towards base. Cal Mateer, British Columbia. Winter/Spring. 2022. 10cm.

'Lily Vegrin'
Smaller snowdrop with slightly inflated spathe and oblong rounded green ovary. Leaves erect to arching, slender, blue-green, paler median line. Outer segments green wash across segment from above apex almost to base, broad white margin. Inner segments green mark above sinus almost to base, narrow white margin. Winter/Spring. 8cm.

'Lilliput'
Small but perfectly formed! Leaves erect, slender, blue-green, paler median line. Outer segments rounded tapering to bluntly pointed apex. Inner segments rounded green 'U'-shaped mark above sinus. Field of Blooms 2022. 6-8cm.

'Limetime'
Elegant looking *G. nivalis* with slender green ovary. Leaves applanate, erect to arching, slender, blue-green. Outer segments, rounded to bluntly pointed apex, clawed with shoulder, lime-green lines from above apex merging at base. Inner segments single darker lime-green mark from above sinus diffusing to base. Usually flowers just before or around New Year.

Limetree'
Slender-looking heavily green-marked flowers with slender green ovary. Leaves slender. Erect to arching. Outer segments slender, slender claw, rounded to bluntly pointed apex, faint lime-green lines above apex to base. Inner segments lime-green mark above sinus diffusing towards base. An Oliver Wyatt selection said to have been found under the Lime tree where Conservative politician Rab Butler drafted his 1944 education reform Act.
15cm.

'Lin'
Neat-looking Hybrid with elegant flowers and rounded green ovary. Outer segments rounded, tapering to bluntly pointed apex. Inner segments rounded green 'U'-shaped mark above sinus. Named for Lin Sales. 2019.

'Lincoln Green'
A green-tipped *G. woronowii* similar to *G.* 'Cider with Rosie' but with darker green markings. Leaves supervolute, broad, splayed, bright green. Outer segments broad, rounded to bluntly pointed apex, dark-green mark above apex. Inner segments broad, dark-green inverted 'U'- shaped mark above sinus. Named as it was found in a Lincolnshire garden centre. Rainbow Farm. 2021.

'Little Angel'
Tiny but very attractive albino *G. nivalis* snowdrop with yellow-green scape, pale-yellow spathe, ovary and pedicel. Leaves applanate, erect, slender, blue-green. Outer segments rounded to pointed apex. Inner segments broad. Very slow to increase. Richard Bashford. 8cm.

'Little Artist'
G. nivalis with small, green-tinged flowers held down within the leaves. Leaves applanate, erect to arching, slender, very green. Outer segments tapering to pointed apex. Inner segments yellow-green inverted 'V'-shaped mark above sinus. Winter/Spring. 10cm.

'Little Beauty'
Neat, small, *G. nivalis* double with more open, outward facing flowers and long, slender green ovary. Leaves applanate, erect, slender, blue-green. Outer segments flaring, tapering to bluntly pointed apex. Inner segments, neat, open ruff, long, slender, green inverted 'V'-shaped mark above sinus towards base. Freddy Van Houtte. Winter/Spring. 10cm.

'Little Bell'
Smaller snowdrop with rounded, 'bell-shaped' green-marked flowers. Outer segments rounded, tapering to bluntly pointed apex, broad green mark above apex. Inner segments green mark above sinus. Winter/Spring. 8cm.

'Little Green Grasshopper'
G. nivalis with slender, green-marked, Poculiform flowers and Scharlockii type split spathe. Leaves applanate, erect, slender, blue-green, paler median line. Outer segments long, slender, incurved to bluntly pointed apex, light-green mark above apex almost to base. Inner segments slightly shorter than outer with green marks above sinus almost to base. Freddy Van Houtte. Winter/Spring. 10cm.

'Little Prince'
Small *G. nivalis* spiky double with very green, upward facing flowers and slender, narrow ovary with petaloid segments. Leaves applanate, erect to arching, slender, blue-green. Outer segments very slender, green marked. Inner segments green marks from above sinus to base. Winter/Spring. 8cm.

'Lola'
G. elwesii x G. nivalis 'Flore Pleno' with rounded, upward facing flowers. Outer segments rounded to bluntly pointed apex. Inner segments broad, neatly packed whorl with green mark apex to base which can be waisted. Found by Phil Cornish in Gloucestershire. c.1990s.

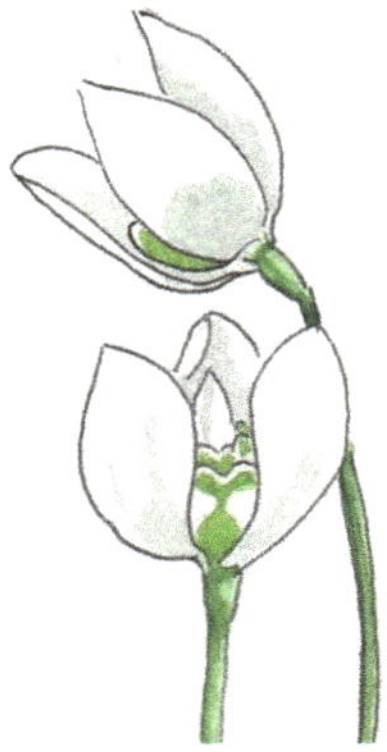

'Lonely Island'
An Inverse Poculiform *G. nivalis*. Leaves applanate, splayed, lanceolate, grey-green. Outer segments spoon-shaped with small green lines above apex. Inner segments rounded with narrow sinus and broad green inverted 'U'-shaped mark above sinus. Named as this snowdrop survived alone when serious fungal infection wiped out all those around it. Krzysztof Ciesielski and his wife near Opie 16-18cm.

'Longbow'
Slender-looking flowers on erect scapes with rounded, light-green ovary and pedicel. Outer segments slender, tapering to pointed apex. Inner segments, slender, elongated green mark above sinus towards base. Ian Christie.

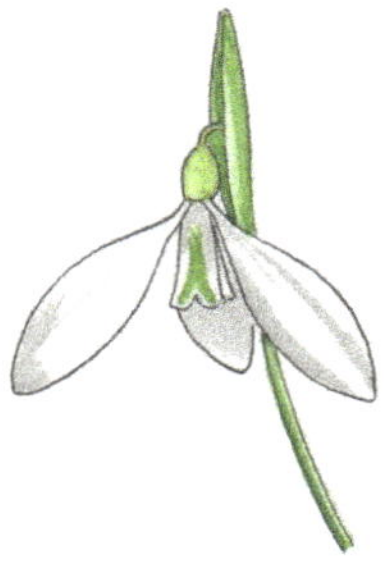

'Longheart'
Large, well-rounded flowers with slender oblong green ovary. Outer segments broad, rounded, clawed, tapering to pointed apex, lightly longitudinally ridged. Inner segments green heart-shaped mark above sinus extending in a narrow line towards base.

'Longwood Green Tip'
Large, weighty-looking, green-marked flowers. Leaves broad, erect, green. Outer segments rounded to pointed apex, bold green marks above apex to one third of segment. Inner segments, broad inverted 'U'-shaped mark to half of segment. Peter Zale at Longwood, USA. 2018. 16cm.

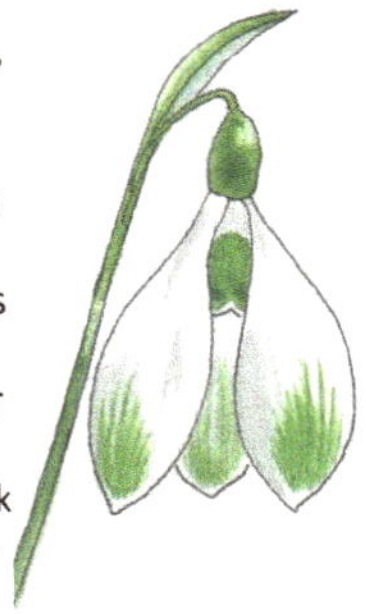

'Look At Me'
Outward-facing double flowers with slender green ovary, throwing additional green-marked segments from ovary. Leaves erect to arching, slender, green. Outer segments slender, rounded to bluntly pointed apex, green marks above apex on lower third of segment. Inner segments green inverted 'V'-shaped mark above sinus. Freddy Van Houtte. Winter/Spring. 14cm.

'Look At Yourself'
G. nivalis with long, slender, green-marked flowers, oblong green ovary, long slender pedicel and split spathe. Leaves applanate, erect, slender, green. Outer segments long, slender, tapering to pointed apex, good green marks above apex towards base. Inner segments good heart-shaped mark above sinus. Freddy Van Houtte. Winter/Spring. 14cm.

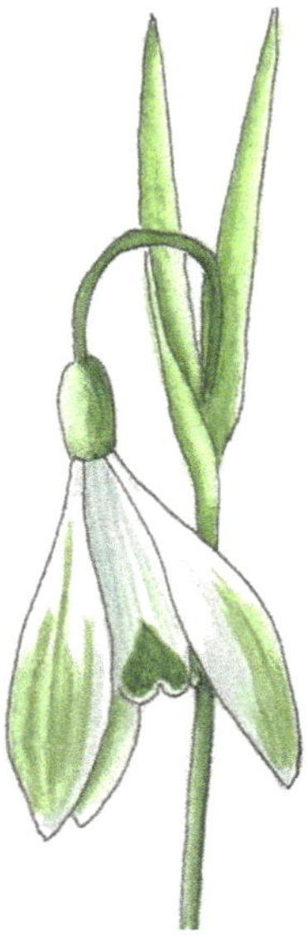

'Louella'
Good. tall, early, yellow-marked x valentinei, a cross between *G.* 'Golden Fleece' and *G.* 'Ecusson D'or' with erect scapes, slender yellow-green pedicel, and yellow ovary. Outer segments broad, flattened, rounded to rounded apex, yellow mark above apex but not always stable. Inner segments yellow-green mark above

'Louise'
Well-marked, late virescent snowdrop with strong green marking suspended from long slender pedicel. Outer segments rounded, tapering to bluntly pointed apex, strong green lines well above apex diffusing at base. Inner segments dark-green mark across segment from above sinus, narrow white margin. Vigorous. March. 12cm.

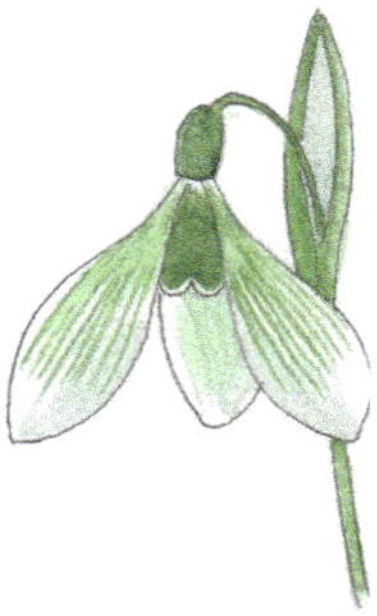

'Magic'
Rounded, green-marked flowers with long slender pedicel giving good movement, and oblong yellow-green ovary. Outer segments broad, flattened, rounded to rounded apex, lightly longitudinally ridged, broad green mark above apex. Inner segments shorter than outer, green mark across segment from above sinus almost to base, narrow white margin. *G. plicatus x G. elwesii*. Seedling from the garden of the late Veronica Cross and named for her lurcher dog.

'Magnetic Pedicel'
Slender-looking flowers on erect scapes with oblong green ovary. Outer segments short claw, slender, tapering to pointed apex. Inner segments broad green horse-shoe shaped mark above sinus.

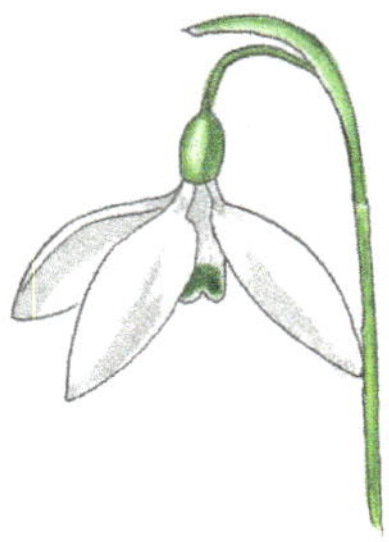

'Maja's Gift'
Rounded white flowers with oblong green ovary. Outer segments rounded, tapering to pointed apex. Inner segments, narrow green inverted 'V'-shaped mark above sinus.

'Margaret Ford'
Slender elongated flowers. Outer segments long, boat-shaped with bluntly pointed apex and light longitudinal ridging. Inner segments green mark from above sinus almost to base, narrow white margin. January/February.

'Margaret Markham'
A *G. nivalis* cultivar which is quite possibly the same as *G.* 'Viridapice' although some growers consider them to be different. Leaves flat to subrevolute margins, erect to arching, slender, blue-green. Outer segments slender, rounded to pointed apex, merging green lines above apex. Inner segments green inverted 'V'-shaped mark above sinus. January/February. 16cm.

'Margaret Mitchell'
Good *G.reginae-olgae* subs. *vernalis* with rounded flowers on erect scapes and rounded, oblong green ovary. Leaves applanate with flat-subrevolute margins, slender, paler median line, short or absent at flowering. Outer segments broad, rounded to bluntly pointed apex. Inner segments broad green inverted 'U'-shaped mark above sinus.

'Marion's Choice'
Striking *G. elwesii* with long slender flowers suspended from angled pedicel, slender curving spathe and long slender dark-green ovary. Leaves supervolute, erect, broad, grey-green. Outer segments long, slender, slender claw, tapering to bluntly pointed apex. Inner segments rounded, inverted green 'V'-shaped mark above sinus.16cm.

'Marion's Surprise'
Very elegant flowers on erect scapes with slender spathe and smaller ovary. Leaves erect. slender, grey-light green. Outer segments beautifully shaped, slender claw with pronounced shoulder, tapering to pointed apex, very lightly longitudinally ridged and textured. Inner segments, light green mark above sinus to three quarters of segment. This snowdrop together with G. 'Marion's Choice' above, probably originated in the garden of the late Marion Saxton, Christchurch, New Zealand, and previously from the small town of Pheasant Point, South Canterbury. Her mother in law specialised in small bulbs and alpines and had an extensive collection of snowdrops which Marion was able to rescue after her mother in law's death, around 50 years ago.

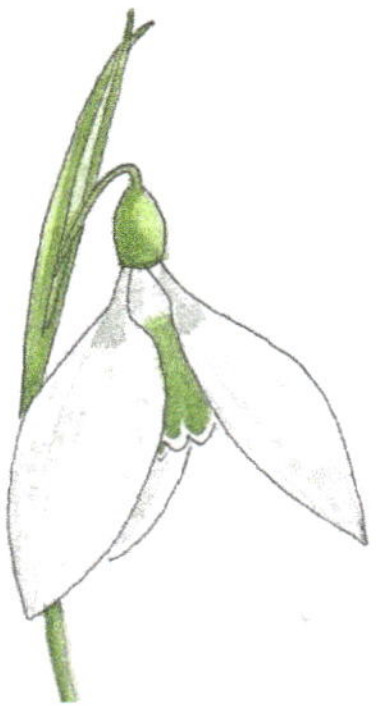

'Marman's Emerald'
Good *G. bursanus* with erect scapes and oblong green ovary. Leaves applanate, medium, blue-green, short at flowering. Outer segments incurved, tapering to pointed apex. Inner segments green mark above sinus almost to base, narrow white margin. Autumn. 12cm.

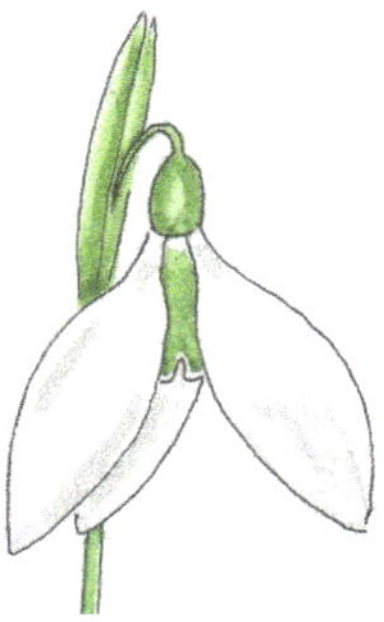

'Marmot Chartreuse'
Weighty looking, well-rounded flowers with yellow markings, yellow-green spathe and ovary. Outer segments very broad, rounded, lightly longitudinally ridged tapering to pointed apex, yellow mark above apex. Inner segments yellow-green mark across segment from above sinus almost to base, narrow white margin. Cal Mateer, British Columbia. 2019. Winter/Spring.

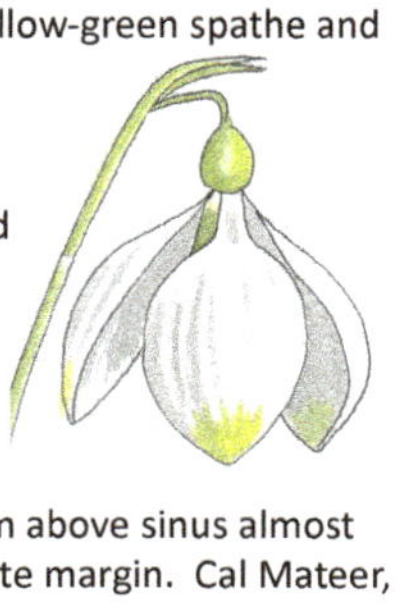

'Marmot Cross'
Rounded, green-marked flowers on erect scapes with rounded green ovary. Outer segments rounded to bluntly pointed apex, longitudinally ridged, broad green mark above apex. Inner segments broad green inverted 'U'-shaped mark above sinus joining smaller oval almost to base. Cal Mateer, British Columbia.

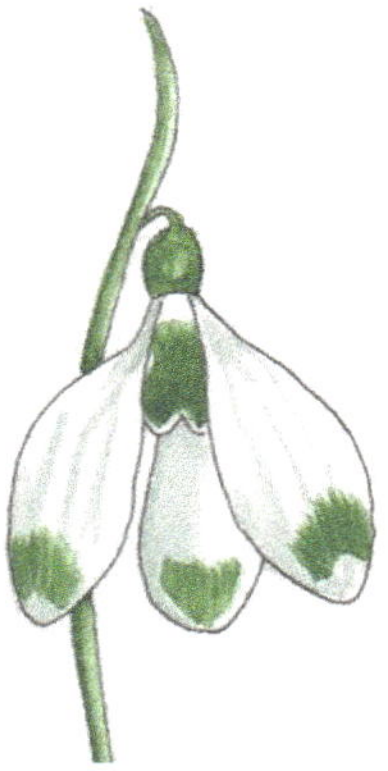

'Marmot Duncan'
Green-tipped *G. elwesii* with erect scapes, slightly inflated spathe and very dark-green ovary. Leaves supervolute, erect to arching, broad, grey-green. Outer segments slender, rounded to pointed apex, small very dark-green marks above apex. Inner segments broad dark-green, almost black, inverted 'V'-shaped mark with upturned ends above sinus joining second broad mark almost to base. Cal Mateer, British Columbia. 2021. Winter/Spring.

'Marmot Flight'
Smaller *G. elwesii* with slender flaring, green-marked segments and rounded green ovary. Leaves supervolute, broad, grey-green. Outer segments slender, incurved, pointed at apex, green mark above apex. Inner segments broad green inverted 'U'-shaped mark above sinus, almost joining second mark towards base. Named as flaring flowers resemble birds in flight. Cal Mateer, British Columbia.

'Marmot Ghost'
Heavily green-marked flowers. Outer segments incurved rounded to pointed apex, merging green lines above apex. Inner segments green mark from above sinus almost to base, longitudinally divided along centre of mark. Cal Mateer, British Columbia.

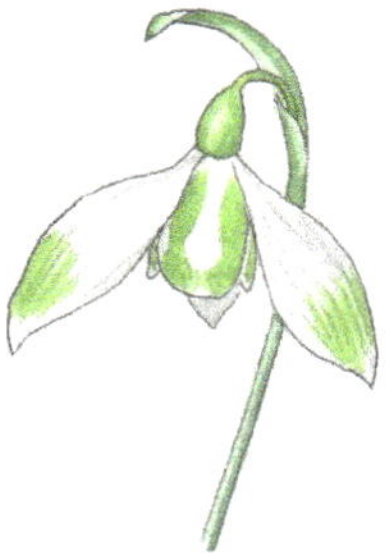

'Marmot Koksilah'
Rounded, green-tipped flowers with rounded green ovary. Outer segments broad, rounded tapering to pointed apex, green mark above apex to one third of segment. Inner segments green mark above sinus almost to base. Cal Mateer, British Columbia.

'Marmot Puckered'
Well-rounded, textured flowers beneath curving spathe and oblong green ovary. Outer segments broad, rounded, lightly longitudinally ridged and puckered, tapering to pointed apex, just a hint of small green marks above apex. Inner segments, broad green mark above sinus to half of segment. Cal Mateer, British Columbia.

'Marmot Sage Grouse'
Attractive green tipped snowdrop with broad, grey-green leaves and large flowers. Outer segments rounded, longitudinally ridged and textured, tapering to bluntly pointed apex, small pale-green lines above apex. Inner segments very dark green heart-shaped mark above sinus merging into oval towards base, narrow white margin. Named as lines on outer segments look like a sage grouse tail. Cal Mateer, British Columbia. 2020. January-February. 15cm.

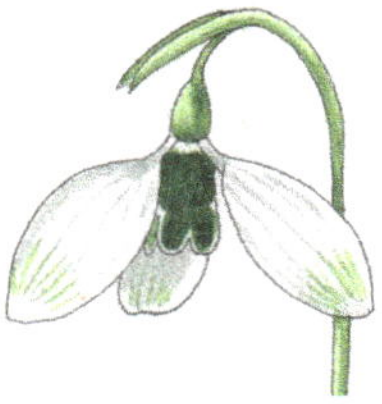

'Marmot Shocker'
Large, long slender flowers with slender, oblong, green ovary. Outer segments slender claw, tapering to pointed apex, longitudinally ridged, green mark above apex. Inner segments very short compared to outer, green heart-shaped mark above small sinus, joining second mark towards base. Cal Mateer, British Columbia. Winter/Spring.

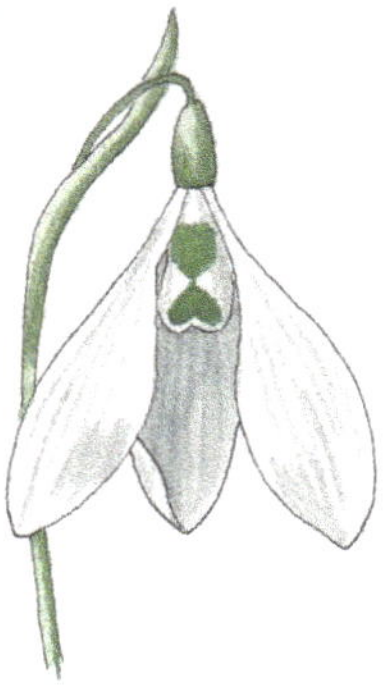

'Marmot Splits'
Elegant looking flowers on erect scapes with rounded green ovary. Leaves broad, erect, grey-green. Outer segments long, flaring, tapering to pointed apex, green mark either side of apex. Inner segments, broad green inverted 'U'-shaped mark above sinus. Cal Mateer, British Columbia. 2019. Winter/Spring. 15cm.

'Marmot Stripe'
Slender-looking, green-marked flowers. Outer segments rounded to pointed apex, green lines above apex to one third of segment. Inner segments green inverted 'V'-shaped mark above sinus joining second mark almost to base. Cal Mateer, British Columbia.

'Marmot Teardrop'
Shorter snowdrop with slender tapering green-marked flowers and oblong green ovary. Leaves erect, broad, grey-green. Outer segments incurved, rounded to bluntly pointed apex, yellow-green mark above apex. Inner segments green mark. Cal Mateer, British Columbia. Winter/Spring. 10cm.

'Marmot Two'
Good *G. elwesii* green-tipped snowdrop producing two flowers. Leaves supervolute, erect, broad, grey-green. Outer segments rounded, tapering to pointed apex. Inner segments green inverted 'V'-shaped mark above sinus, second mark towards base. Cal Mateer, British Columbia.

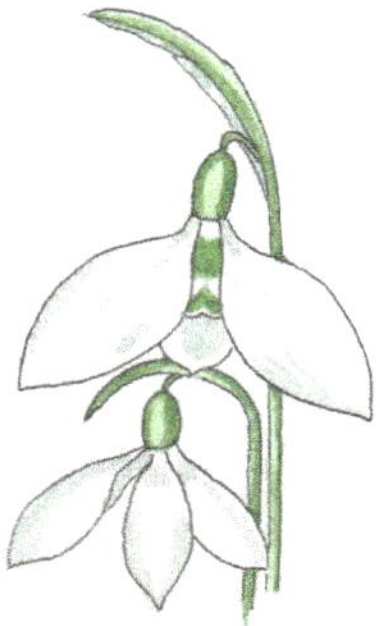

'Maryse's Double'

A snowdrop that always has upward-facing, green-marked flowers and small triangular green ovary. Outer segments rounded, tapering to pointed apex, green mark above apex. Inner segment ruff, green marks above sinus. Found by Oliver Vico in a private garden near Normandy, France. The snowdrop is named for the garden owner who kindly gave him bulbs.

'Marzenbecher'

- Spiky with outward facing green-marked flowers on erect scapes with short spathe. Leaves erect to splayed, slender, green, paler median line. Segments roughly of equal length, narrow, tapering to pointed apex, green mark above apex to half of segment. 'Marzenbecher' has been used to reference the spring snowflake, Leucojum = cup-shaped. So presumably named here for the snowdrop's cup shaped flowers. Winter/Spring. 2021. 14cm.

'Maulsden'

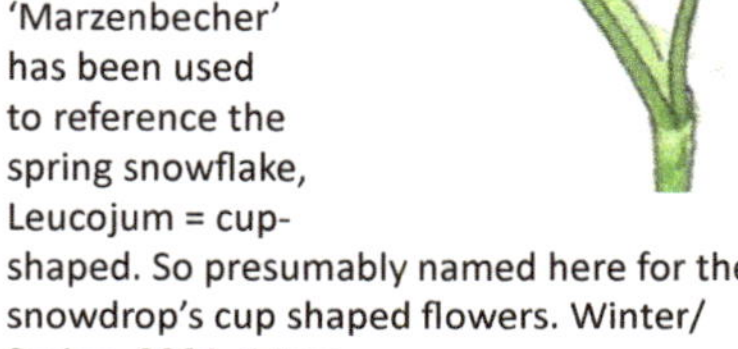

Well rounded, lightly textured flowers on erect scapes with oblong, rounded ovary. Outer segments broad, rounded, short claw, rounded at apex. Inner segments double green mark, inverted 'V'-shaped mark above sinus, second shield-shaped mark towards base.

'Medena'

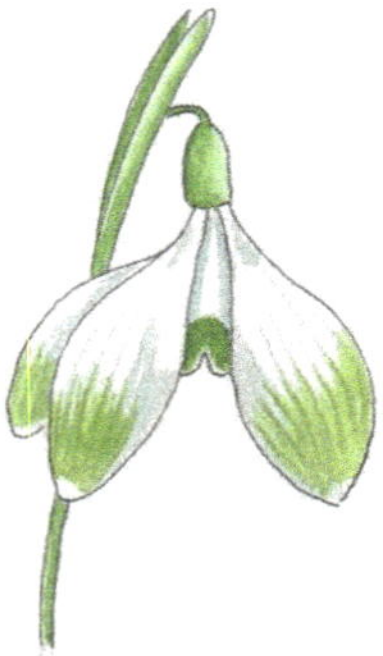

G. nivalis with beautifully green-marked flowers on erect scapes with oblong green ovary. Leaves applanate, erect to arching, slender, blue-green, paler median line. Outer segments rounded and tapering to bluntly-pointed apex, green lines above apex to half of segment. Inner segments green heart-shaped mark above sinus. Angelina Petrisevac, Croatia. January/February. 15cm.

'Melody in Green'

Attractively marked virescent snowdrop. Leaves applanate, slender, erect to splayed, blue-green, paler median line. Outer segments rounded, tapering to bluntly pointed apex, pale green wash across segment from above apex almost to base. Inner segments mid-green, horse- shoe shaped mark above sinus diffusing into yellow-green on basal side. Gert Jan van der Kolke. 2021. Winter/Spring. 16cm.

'Meltsas'

Estonian bird series of snowdrops. (Meltsas = Green Woodpecker - *Picus viridis*). Beautiful well-rounded, double flowers with tightly packed, green-marked inner segments, slightly inflated spathe and short pedicel with small triangular ovary. Outer segments broad rounded, rounded at apex, longitudinally ridged, merging green lines from above apex. Inner segments tight, neat, well-formed ruff heavily marked green from above sinus to base. Taavi Tuulik, Estonia. 2020.

'Merlin Bonds Form'

More rounded flowers than *G.* 'Merlin' with rounded green ovary. Outer segments broad, rounded, tapering to pointed apex. Inner segments green mark from above sinus almost to base, narrow white margin.

'Merlins Herbstgefoige'

Elegant-looking *G. peshmenii* with well-shaped flowers. Leaves applanate, slender, grey-green, usually absent or small at flowering. Outer segments long tapering to pointed apex. Inner segments, broad, rounded, green heart-shaped mark above sinus fading on basal side. Autumn.

'Meteor'

Beautifully shaped, large flowers with long arching pedicel and slender green ovary. Outer segments broad, rounded, short claw, rounded at apex, small green lines above apex. inner segment mark filled in inverted 'U'-shaped mark above sinus.
Richard Bashford, Woodchippings

'Mett'

Attractive balloon-like double flowers. Outer segments well rounded, tapering to pointed apex, merging green lines above apex to half of segment Full inner ruff of heavily green-marked segments.

'Mex'

Green-marked flowers with oblong, rounded green ovary. Outer segments clawed, rounded to bluntly pointed apex, good green lines above apex to half of segment. Inner segments solid mid-green mark across segment from above sinus to base.

'Midori'

G. elwesii var. *monostictus* with slender, rounded, green-marked flowers. Leaves supervolute, broad, erect, green-grey. Outer segments short claw, rounded to bluntly pointed apex, merging green lines above apex to three-quarters of segment. Inner segment green mark above sinus almost to base, narrow white margin.

'Mila'
Neat looking *G. nivalis* on erect scapes with small green ovary. Outer segments rounded to bluntly pointed apex, variable small green lines above apex. Inner segments slender with dark-green horseshoe shaped mark above sinus diffusing on basal side.

'Milch und Honig' (Milk and Honey)
Slender, rounded, yellow marked flowers. Leaves erect, slender, blue-green. Outer segments slender, rounded to pointed apex. Inner segments slender, small inverted yellow 'V'-shaped mark above sinus, second larger yellow mark towards base. Hagen Engelmann, Germany. 2019. Winter/Spring. 15cm.'

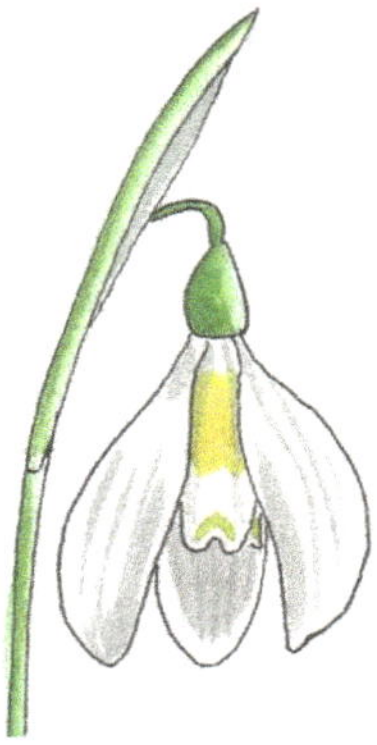

'Minerali Bani'
Large snowdrop raised from seed originating in Bulgaria. Paul Barney, Edulis Nursery. 2020.

'Mini Poculiformis'
Very tiny *G. nivalis* Poculiform snowdrop on short scapes with small green ovary and rounded flowers. Leaves applanate, erect, slender, blue-green. Segments all roughly of equal length, white, rounded to bluntly pointed apex. Josse Bavcon, University Botanic Garden Ljubljana. Slovenia. Winter/Spring. 6cm.

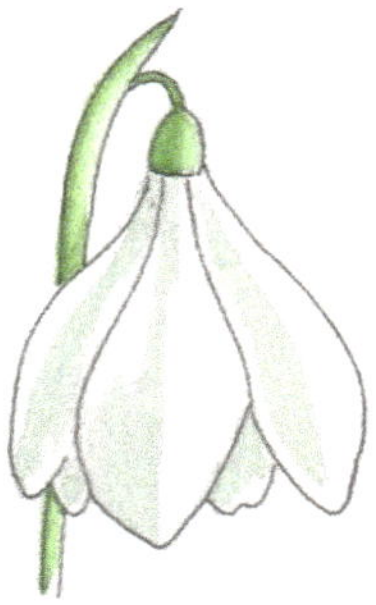

'Minions Smiling'
Small *G. nivalis* with slender flowers and oblong green ovary. Leaves applanate, erect, slender, blue-green. Outer segments slender, tapering to pointed apex. Inner segments slender, green inverted 'U'-shaped mark above sinus. Found in a wood near Zagreb. 8cm.

'Miss Tilly'
An early-flowering *G. reginae-olgae* with slender flowers and slender, oblong green ovary. Leaves applanate, slender, grey-green. Outer segments slender, tapering to pointed apex, lightly textured. Inner segments rounded, green inverted 'U'-shaped mark above sinus. Edgwood Gardens, USA. Autumn/Winter.

'Misty Jewel'
Autumn flowering *G. elwesii* with well rounded flowers tapering to pointed apex. Leaves supervolute, erect, broad, grey-green. Outer segments rounded, tapering to bluntly pointed apex, green lines above apex to one third of segment. Inner segments heart-shaped green mark above sinus merging into inverted heart-shaped mark towards base. October. Ruben Billiet.

'Mmm'
G. elwesii with rounded flowers and oblong yellow-green ovary. Leaves supervolute, erect, broad, grey-green. Outer segments clawed, rounded, tapering to bluntly pointed apex. Inner segments inverted green 'U'-shaped mark above sinus.

'Moby Dick'
Good hybrid with large flowers suspended from long slender pedicel, although the plant is not exactly whale size! Leaves erect, glaucous. Outer segments rounded to bluntly pointed apex. Inner segments, spreading, green inverted 'V'-shaped mark with rounded ends above sinus, two pale 'eye' spots towards base, attempting a smile perhaps? 2021. Winter/Spring.

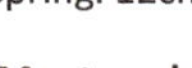

'Money'
G. nivalis Poculiform Scharlockii snowdrop with rounded, heavily green marked, upward facing flowers and broad Scharlockii split spathe. Leaves applanate, erect. slender, green. All segments roughly of equal length, rounded, rounded at apex, heavily marked green apex to base. Freddy Van Houtte. Winter/Spring. 12cm.

'Montrose'
Autumn flowering *G. elwesii* var. *monostictus* originally from Montrose Gardens, North Carolina, USA. and thought to have been planted by Nancy Goodwin in the late 1980s. Slender pedicel with triangular shaped green ovary. Leaves broad, green, reflexing and short at flowering. Outer segments broad, flattened, rounded at apex. Inner segments green inverted 'U'-shaped mark above sinus and distinct orange shadow at base. Autumn/Winter. 8-10cm.

'Moonlight Shadow'
Green-marked *G. nivalis* with oblong green ovary. Leaves. applanate. erect. slender, blue-green. Outer segments tapering to pointed apex, green mark above apex to three-quarters of segment. Inner segments solid green mark above sinus to base, narrow white margin. Angelina Petrisevac. 2018

'Moonshine'
Slender-looking flowers with unusual pale yellow-green scape, spathe and pedicel and yellow ovary. Leaves erect, slender, pale grey-green. Outer segments slender, incurved, tapering to pointed apex. Inner segments yellow mark above sinus diffusing towards base. Richard Bashford. 2020. Winter/Spring.10-15cm.

'Moose Bear'
Semi-poculiform snowdrop with four green marks on the inner segments. Outer segments rounded to bluntly pointed apex, light- green tips. Inner segments slightly shorter than outer, green mark either side of apex and a second mark each side of segment at base. Cal Matter. British Columbia. Winter/Spring. 2021.

'More Than A Feeling'
G. nivalis Sharlockii Poculiform snowdrop with green marked flowers, split spathe and long, slender pedicel. Leaves applanate, erect to arching, slender, blue-green, paler median line. Segments slender, rounded to bluntly pointed apex, green mark above apex tapering up to half of segment. Freddy Van Houtte. Winter/Spring. 12cm.

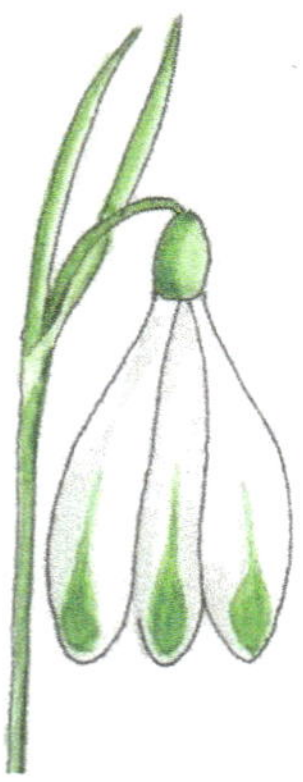

'Moses Autumn Jewel'
Early-flowering snowdrop with long, slender flowers. Leaves broad, arching, grey-green. Outer segments long, slender, pointed at apex. Inner segments green mark above sinus almost to base, can split into two elongated marks. September/October. 10-12cm.

'Motabile'
An unusual four-by-four *G. gracilis* Inverse Poculiform snowdrop with small, green-marked flowers and large, rounded green ovary. Leaves applanate, erect, slender, slightly twisted. Outer segments broad, rounded at apex, green mark above apex and second green mark towards base. Inner segments green mark above sinus.

'Mr Camoflage'
Rounded flowers with good green marks and rounded triangular ovary. Outer segments rounded to bluntly pointed apex, small green lines above apex. Inner segments broad green mark above sinus, diffusing slightly on basal side. Marcel Kuschel. 2022.

'Mr Courage's Early'
Weighty-looking *G. elwesii* var. *monostictus* with large flowers on short, erect scapes. Leaves erect, broad, grey-green, short at flowering. Outer segments long, boat-shaped, rounded at apex, longitudinally ridged. Inner segments slightly variable marks from two merging green ovals above sinus to an inverted 'U'-shaped mark. Autumn/Winter. 8-10cm.

'Mr Frayling's Double'
Double *G. nivalis* with rounded flowers. Leaves applanate, erect. slender, blue-green. Outer segments rounded, spreading, rounded to bluntly pointed apex. Inner segments full ruff with green mark above sinus. David Bromley, Shropshire. January/February. 15cm.

'Mr Peggoty'
G. elwesii with rounded white flowers and oblong green ovary. Leaves supervolute, erect, broad, grey-green. Outer segments broad, rounded, tapering to bluntly pointed apex, lightly longitudinally ridged. Inner segments green Chinese-bridge shaped mark above sinus.

'Mr Positive'
Heavily green marked flowers with oblong green ovary. Outer segments long, slender, tapering to pointed apex, green lines from one third above apex, running to base. Inner segments green mark across segment above sinus, diffusing at base, narrow white margin.

'Mr T. Mackie'
Another in the Inverse Poculiform group of snowdrops with good green marks. With *G.* 'South Hayes' in its parentage it is taller and stronger growing than that snowdrop with more arching stems. Leaves, wavy and puckered. Outer segments broad, flattened, rounded at apex, double green mark, one towards base and a second above apex, which can turn more golden/bronze in very mature flowers. Inner segments heavily marked green. Good, strong grower. John Holland. 28-30cm.

'Mr Taylor'
Early to mid-season hybrid convolute with elegantly shaped flowers and good green marks beneath oblong, rounded green ovary. Leaves broad, erect, convolute, grey-green. Outer segments broad, rounded to pointed apex, short merging green lines above apex. Inner segments mid-green mark above sinus almost to base, narrow white margin. Avon Bulbs from Veronica Cross's garden. 2019. Autumn/Winter.

'Mutant'
Small, neat *G. woronowii* Inverse Poculiform with small flowers and large oblong yellow-green ovary. Leaves supervolute, broad, ridged, grey-green. Segments rounded, rounded at base with small sinus and light green dot either side of sinus. Richard Bashford. January. 10cm.

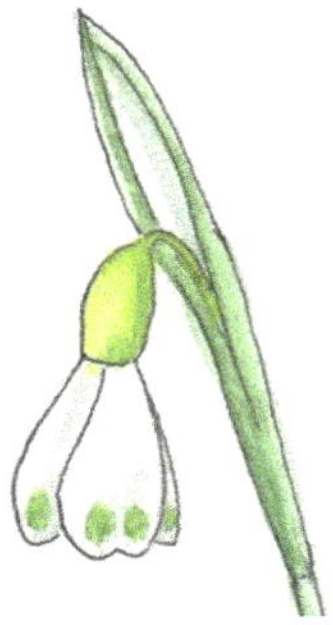

'Neat Green'
Attractive smaller *G. nivalis* cultivar with very green distinctive foliage, similar to G. 'Pastures Green'. Leaves applanate, erect to arching, slender, bright shiny green. Outer segments slender, rounded, tapering to pointed apex. Inner segments narrow inverted green 'V'-shaped mark above sinus. Seedling from Michael and Ann Broadhurst, Rainbow Farm and found in a large colony of standard *G. nivalis* in Cambridgeshire. Good and bulks up well. January/February. 10-12cm.

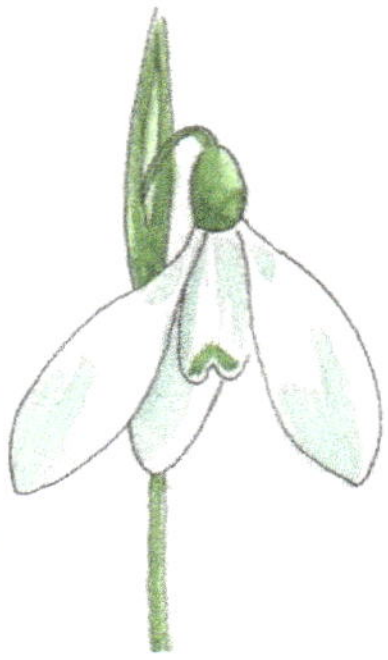

'Nellie Brinsley's Double'
Full double *G. nivalis* with well proportioned, very rounded flowers, similar looking to *G.* 'Pusey Green Tips'. Leaves applanate, erect, to arching, slender, blue-green. Outer segments, rounded, incurved, tapering to pointed apex, green mark above apex. Inner segments untidy variable ruff with inverted green 'V'-shaped mark above sinus. January/February. 15cm.

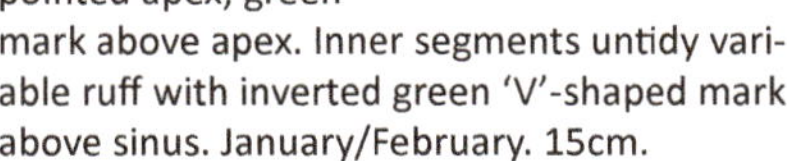

'Nellie-e-Belle'
Beautiful neat looking *G. nivalis* Inverse Poculiform with long slender pedicel. Leaves applanate, erect slender, blue-green. Outer segments flattened, rounded to rounded apex, green mark above apex. Inner segments slightly shorter than outer, green mark above sinus. John Alpassa and named for his wife. 2021. Winter/Spring. 16-18cm.

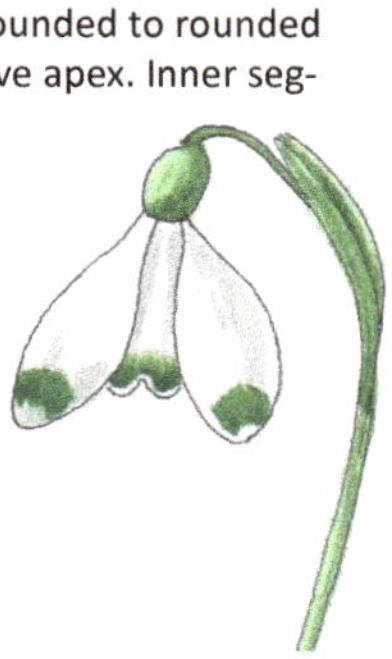

'Nellie's Birthday'
G. elwesii var. *monostictus* with strong green marking on erect scapes with slender, triangular green ovary. Leaves supervolute, broad, erect, grey-green. Outer segments broad, rounded to pointed apex, strong mid-green mark above apex on lower third of segment. Inner segments green mark across segment from above sinus almost to base, narrow white margin. Hagen Engelmann and named for a friend as it usually flowers for her birthday.

'Neon Kiss'
Handsome rounded flowers with oblong green ovary. Outer segments broad, short claw, rounded, tapering to pointed apex. Inner segments Chinese-bridge shaped mark above sinus.

'Nessie'
Yellow marked *G. nivalis* with long slender buds. Leaves applanate, erect, slender, blue-green. Outer segments slender, tapering to bluntly pointed apex. Inner segments yellow inverted 'U'-shaped mark above sinus with paler yellow shading across segment. Freddy Van Houtte. Winter/Spring. 10-15cm.

'New Art'
Slender-looking, rounded, green-marked flowers with oblong green ovary. Outer segments slender, rounded tapering to bluntly pointed apex, broad green mark above apex to lower third. Inner segments slightly shorter than outer, broad green mark above sinus.

'Norma'
Rounded flowers with rounded green ovary. Outer segments short claw, rounded to bluntly pointed apex, lightly longitudinally ridged. Inner segments broad mid-green mark across segment from above sinus to three quarters of segment, narrow white margin.

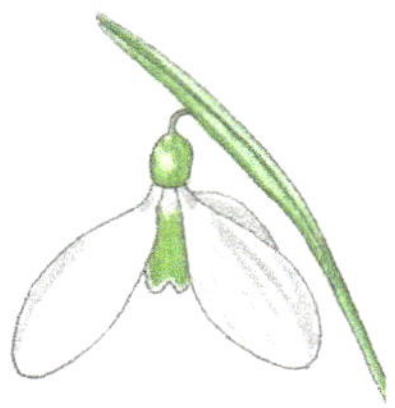

'North Green Wasp'
G. elwesii with oblong green ovary and rounded segments. Leaves supervolute, erect to arching, broad, grey-green, short at flowering. Outer segments clawed, long, slender, rounded, incurved, tapering to bluntly pointed apex. Inner segments green mark above sinus almost to base, narrow white margin. North Green Snowdrops.

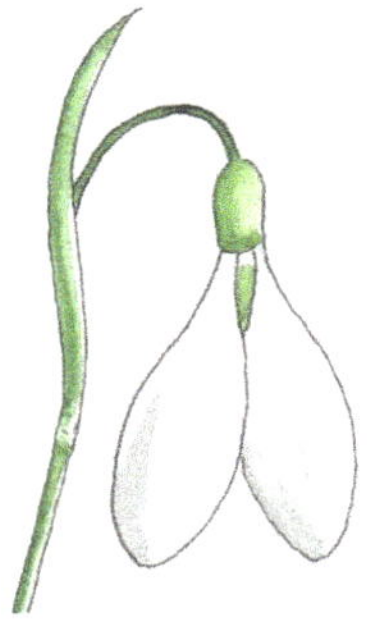

'Nutt's Early'
Attractive early *G. elwesii* var. *monostictus* with long, pendant flowers. Leaves supervolute, erect, broad, grey-green. Outer segments long, rounded tapering to bluntly pointed apex. Inner segments green inverted heart-shaped mark above sinus merging into larger mark almost to base. From the garden of Galanthophile the late Richard Nutt who is said to have given it away to friends at one of his snowdrop lunches. Bulks up well. Early, usually before Christmas. Autumn/Winter. 16cm.

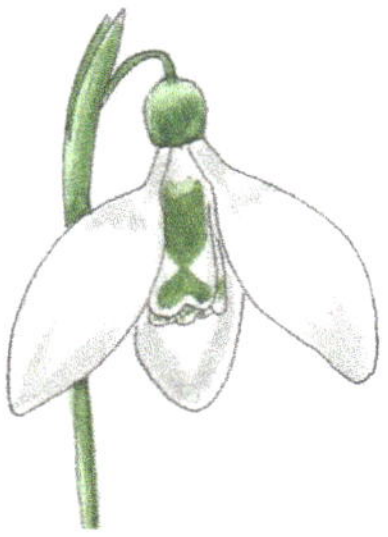

'Odd Sharlock'
A Sharlockii snowdrop with long, strongly angled pedicel, long split spathe and good green-marked flowers. Leaves erect to splayed, slender, green-blue, paler median line. Outer segments rounded to pointed apex, strong green mark above apex. Inner segments dark-green broad inverted 'V'- shaped mark above sinus. Originally from the late Margaret Owen's garden in Shropshire, she recognised this as being distinct from the usual *G. nivalis* Scharlockii and therefore named it. 2020. Winter/Spring. 8-10cm.

'Of'

Distinctive-looking *G woronowii* clone with attractive foliage and well-proportioned flowers. Leaves supervolute, long, strap-like, shiny green. Turkey.

'Okidoki'

Small virescent snowdrop with good green marking. Leaves slender, erect, blue-green. Outer segments merging green lines above apex to half of segment. Inner segment green mark above sinus diffusing towards base. 10cm.

'Olana'

Lower growing, autumn/winter flowering *G. plicatus* with oblong mid-green ovary. Leaves explicative, reflexed, grey-green. Outer segments slender, rounded, tapering to bluntly pointed apex, merging green lines from above apex to base. Inner segments single mid-green mark above sinus almost to base, narrow white margin November-January. 8cm.

'Old Kite'

Interesting, distinctive green-tipped Poculiform snowdrop with long green divided spathe. Leaves applanate, erect, slender, blue-green. Segments of equal length, rounded, rounded at apex with good yellow-green marks spreading upwards from above sinus. From the late Margaret Owen, Acton Pigot, Shropshire and labelled by her as '*Galanthus nivalis* Virescens - 'Doing a Kite' it is not a virescent snowdrop and has been confirmed it originated with Johan Mens.

'Olivia Newton John'

Slender green marked flowers with oblong green ovary. Outer segments long, slender, incurved, tapering to bluntly pointed apex, mid-green wash above apex to three quarters of segment. Inner segments broad mid-green mark above sinus. Oliver Vico.

'Orange Star'

A strange-looking *G. nivalis* with very open flowers in a loose arrangement of spreading segments, some tinged orange and revealing golden anthers. Leaves applanate, erect, slender, blue-green. Outer segments long, very slender, tapering to pointed apex. Inner segments small, slender, tinged orange giving the appearance of anthers. Found by Oliver Vico and Ruben Billiet in Northern France.

'Orion'

A simple-looking, tall *G elwesii* with slender flowers. Leaves supervolute, erect, medium, grey-green-glaucous. Outer segments long, clawed, slender, tapering to pointed apex. Inner segments broad green inverted 'U'- to 'V'-shaped mark above sinus. December-January. 16cm.

'Ostergras'

Attractive Poculiform with flaring green-tipped flowers and slender green ovary. Leaves erect. slender, blue-green. Segments all of equal length, slender, tapering to pointed apex, green mark above apex on lower quarter. Hagen Engelmann, Germany. 2019. Winter/Spring.

'Otterhauw'

Recently introduced and rare snowdrop throwing extra segments from the ovary. 2020.

'Paleface'

A *G nivalis* in the *G.* 'Norfolk Blond' ilk with long, slender yellow-green ovary. Leaves applanate, erect, slender, blue-green. Outer segments long, incurved, pointed at apex. Inner segments pale yellow mark above sinus. Found in Bradenham Churchyard, Norfolk. 2018.

'Papgeno'

Small, new and charming little snowdrop. Leaves slender, erect, blue-green, paler median line. Outer segments rounded, tapering to bluntly pointed apex, light green merging lines above white apex to two thirds of segment. Inner segments mid-green mark above sinus diffusing towards base. Gert-Jan van der Kolk found in the same wood as *G. nivalis* 'Green Tear', and although similar in appearance, this snowdrop is half the size.

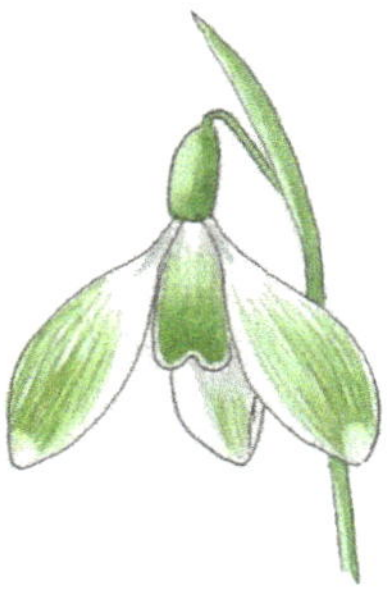

'Parakeet'

A very quirky and interesting *G. elwesii* Inverse Poculiform with extremely variable flowers. Leaves, supervolute, broad, very long up to 30cm. Variable segments from six very green inner segments or 5 inners and one outer, or four inners and two-three outer segments. Originally from Kencot.

'Paranoid'

G. nivalis with very small, green-marked flowers and long Sharlockii type split spathe. Leaves applanate, erect to arching, slender, blue-green. Outer segments incurved, rounded to bluntly pointed apex, green mark above apes. Inner segments green inverted 'W'-shaped mark above sinus diffusing towards base. Freddy Van Houtte. 2021. Winter/Spring.

'Parfetta'
Elegant small *G. nivalis* Poculiform flowers on erect, long slender pedicel with small green ovary and long split spathe. Leaves applanate, erect to arching, slender, blue-green. All segments of equal length and marked green above apex. Johan Mens. 8-10cm.

'Party Dress'
Well-shaped inverse Poculiform snowdrop with good green marks. Outer segments broad, flattened rounded to rounded apex, lightly longitudinally ridged, broad green mark above apex, small paler mark towards base. Inner segments green mark across segment, narrow white margin. Richard Bashford. 2021.

'Pear Bear'
Poculiform with rounded droplets of beautifully pear-shaped white buds, opening into large, pure white, six equal segmented spreading flowers with small green ovary and light longitudinal ridging. Leaves erect, blue-green, reflexed outwards at tip. Cal Mateer, British Columbia. 2022. 18cm.

'Peg's Double'
Loose, rounded, double flowers with small, slender, oblong green ovary. Outer segments rounded, incurved, tapering to pointed apex. Inner segments, loose ruff with small, variable green mark above small sinus.

'Peppermint Candy'
G. nivalis with rounded flowers and strong green marking. Leaves applanate, erect, slender, blue-green, paler median line. Outer segments broad, rounded to bluntly pointed apex, strong green lines above apex to three quarters of segment. Inner segments broad green inverted 'U'-shaped mark above sinus. Snowdropfevers. Winter/Spring. 16cm.

'Pepys' Syn 'Not Sparkler'
A mysterious, weighty-looking *G. elwesii* with very white flowers from the late Margaret Owen with an incorrect label *G.* 'Sparkler'. Leaves supervolute, erect, blue-green. Outer segments large, weighty, rounded, tapering to pointed apex. Inner segments quite small with narrow green mark above sinus and two paler yellow-green 'eye' dots towards base. The late Margaret Owen's garden, Acton Pigot, Shropshire.

'Peter Davis' Syn PHD 33463.
Recently re-named so included again here. Beautifully proportioned *G. plicatus* with large, well-formed flowers. Leaves explicative, erect to arching, grey-green, paler median line. Outer segments brilliant white, long, rounded, longitudinally ridged, light goffering at base, tapering to bluntly pointed apex. Inner segments, clear green heart-shaped mark above sinus bleeding slightly on basal edge. Good fragrance. Discovered by Dr. Peter Hadland Davis and said to be from seed originally collected in Georgia in the 1970s. This was later amended to seed collected in the Crimea with the original collection number of PHD 33643 which was inaccurate and should have read PHD 33463. Specimens were passed to Kew and Leningrad Botanic Gardens and the snowdrop probably came into circulation via Chris Brickell who worked at Kew at the time. Thanks to Alan Briggs, Margaret Thorne and Tim Harberd for their research in clarifying the information on this snowdrop.

'Petra'
Heavily green marked flowers with long rounded green ovary. Leaves erect to splayed, slender, grey-green. Outer segments rounded to bluntly pointed apex, heavily marked green from above apex to base. Inner segments green mark across segment from above sinus, diffusing at base, narrow white margin. Angelina Petrisevac and named for Petra Wagner.

'Petrich'
Seed originally from Bulgaria and later chipped giving a crossover species. Paul Barney, Edulis Nursery. 2020.

'Phoebe'
Large slender, green-marked flowers on slender pedicel with slightly inflated spathe and oblong green ovary. Outer segments broad, tapering to bluntly pointed apex, light-green lines above apex to three quarters of segment. Inner segments broad dark-green mark above sinus diffusing towards base with two darker 'eye' spots. Richard Bashford.

'Pik Dame'
G. nivalis with long, slender flowers, arching pedicel and slender green ovary. Leaves applanate, slender, erect, blue-green. Outer segments long, slender, tapering to pointed apex. Inner segments inverted green 'U'-shaped mark above sinus, bleeding into two paler merging marks towards base.

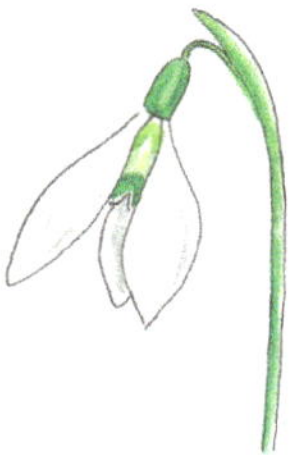

'Pil Pil'
Well-shaped flowers on erect scapes with slender pedicel and slender rounded ovary. Outer segments rounded, tapering to pointed apex. Inner segments green mark from above small sinus almost to base, narrow white margin. Winter.

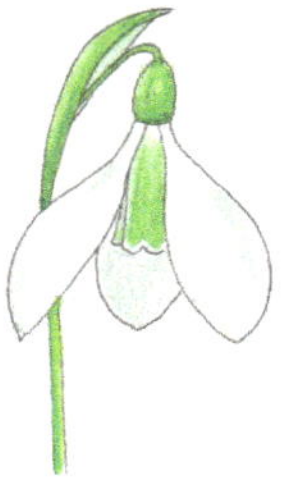

'Pingpoc'
Small quirky rounded flowers with varying segments beneath rounded yellow-green ovary. Outer segments rounded, lightly longitudinally ridged with green mark above apex on broader segments, pale yellow-green mark on smaller segments.'

'Piping Plover'
G. elwesii with slender flowers, flaring, green marked segments and long, slender green ovary. Leaves supervolute, medium, erect, blue-green. Outer segments slender, flaring upwards, green mark above apex. Inner segments long, slender with green, inverted 'U'-shaped mark above sinus.

'Praying Hands'
Rounded flowers likened to hands clasped in prayer, hence name. Outer segments rounded, incurved to bluntly pointed apex. Inner segments mid-green mark above sinus almost to base.

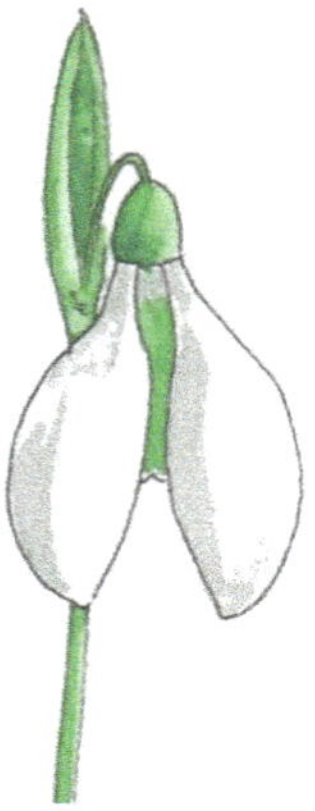

'Priscilla's Green'
Erect scapes with long, rounded ovary. Outer segments clawed, rounded to bluntly pointed apex, green mark above apex. Inner segments, green mark above sinus, almost to base. Winter/Spring.

'Puffin'
Smaller plants with large, rounded, textured flowers, slender pedicel and oblong green ovary. Leaves erect to arching, slender, green. Outer segments broad, rounded, lightly longitudinally ridged and textured, tapering to bluntly pointed apex. Inner segments narrow green inverted 'V'- shaped mark above sinus and second larger green mark towards base. Strong and vigorous, bulks up well. Hilary and Hugh Purkiss, Gloucestershire. Winter/Spring. 12cm.

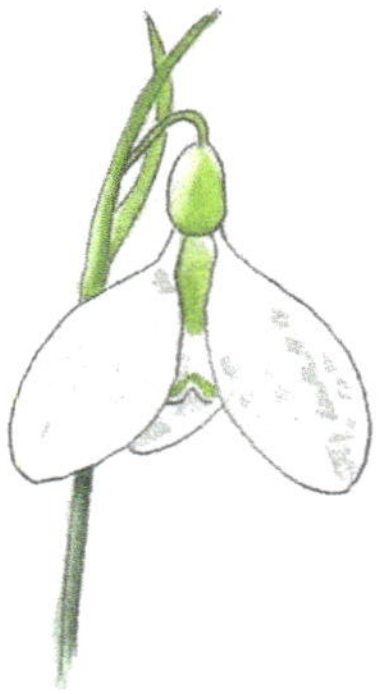

'Pumpkin'
Large, rounded, weighty-looking, well-textured flowers. Leaves erect to splayed, slender, grey-green, paler median line. Outer segments broad, rounded to bluntly pointed apex, longitudinally ridged and textured. Inner segments narrow green inverted 'U'-shaped mark above sinus. Richard Meijndert. 2018. 14cm.

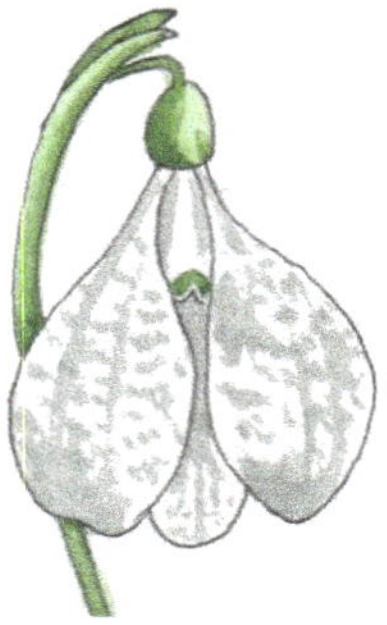

'Punky'
Quirky looking snowdrop with slender, upward facing green flowers held at an erect angle between Scharlockii type split spathe. Outer segments long, slender, rounded at apex, heavily marked green apex to base. Inner segments small, rounded, creamy-green. Oliver Vico. 2022.

'Quasimodo'
Very short *G. elwesii* with odd looking, unpredictable flowers which can include a white spathe. Leaves supervolute, erect, broad, grey-green. Outer segments variable, some slender, some broader, tapering to pointed apex, variable green mark above apex. Inner segments slightly more normal looking with green mark above apex, sinus absent or very small. Helen Squires garden, Oxfordshire. 12cm.

'Rachel Mahaffy'
Early *G. reginae-olgae* with delicate looking flowers on erect scapes with triangular green ovary. Outer segments slender tapering to pointed apex. Inner segments slender green heart-shaped mark above deep sinus. September/October.

'Rainbow Alcyone'
G. elwesii with slender looking flowers. Leaves supervolute, erect, broad, grey-green. Outer segments, slender, flaring, green crescent shaped mark above apex. Inner segments, green inverted 'U'-shaped mark above sinus. Named for goddess of the winter solstice. Vigorous and bulks up well. Rainbow Farm. 2021. Early, usually by December.

'Rainbow Balthazar'
Good, vigorous early flowering hybrid seedling with well rounded flowers and green ovary. Outer segments broad, rounded to rounded apex. Inner segments green mark above sinus. The second of the Rainbow Farm 'Three Kings' series. 2021. Early around Christmas

'Rainbow Bealvi'
Unusual, good, green-marked *G. elwesii* with very erect spathe and long triangular green ovary. Leaves supervolute, broad, erect to arching, grey-green. Outer segments long, slender, rounded to bluntly pointed apex, good green marks above apex between one third and one half of segment. Inner segments, very short, about one third, compared to outer with broad dark-green inverted 'U'-shaped mark above sinus. The second in the Rainbow 'Green Goddess series' and named for the goddess of fertility of plants and animals and the festival of the winter solstice, Rainbow Farm, 2021. Early, usually flowering pre-Christmas.

'Rainbow Caspar'
An attractive large flowered hybrid snowdrop with long, rounded light-green ovary. Outer segments rounded, tapering to bluntly pointed apex. Inner segments broad with dark-green inverted 'V'-shaped mark above sinus and second paler green mark towards base. Usually two scapes on mature bulbs. Third in the Rainbow Farm 'Three Kings' series. 2021. Early usually flowering near Christmas.

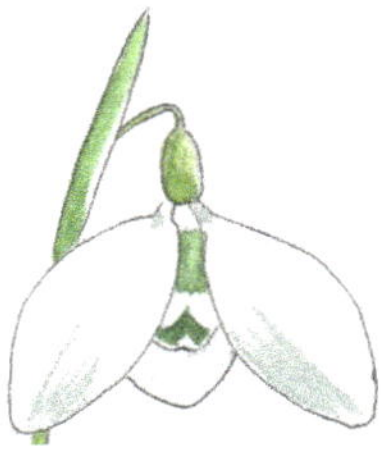

'Rainbow Christmas Fairy'
Smaller snowdrop, possibly a hybrid, with long slender green ovary. Outer segments flaring tapering to bluntly pointed apex. Inner segments green inverted 'U'-shaped mark above sinus, two paler elongated eye marks towards base running towards apex. Rainbow Farm. Flowering around Christmas. 10cm.

'Rainbow Golden Girl'
A good, strong Hybrid with slender green ovary and yellow marking. Outer segments slender, rounded to bluntly pointed apex. Inner segments unusual yellow marking above sinus diffusing towards base forming an 'X'-shaped mark. Good grower although can show colour change some years. Rainbow Farm.

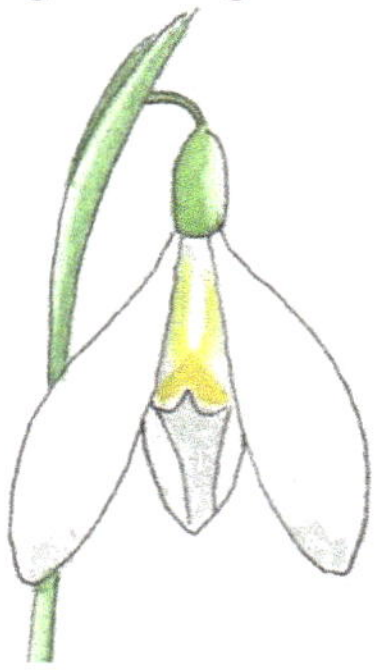

'Rainbow Gold Top'
Smaller, yellow-marked, late flowering Hybrid with yellow pedicel and yellow-green ovary, usually small yellow patch where pedicel joins ovary. Leaves splayed, blue-green. Outer segments long, rounded to bluntly pointed apex. Inner segments golden-yellow inverted 'V'-shaped mark above sinus. Strong growing and later flowering than other Rainbow Farm yellows. Rainbow Farm. 2021. February/March often well into March.

'Rainbow Green Gracie'
Smaller, dainty snowdrop with narrow green ovary, classed as either a *G. gracilis* or hybrid. Leaves applanate, slender, very twisted as in *G. gracilis*. Outer segments boat shaped with small green lines above apex. Inner segment mark also reminiscent of *G. gracilis*. Rainbow Farm. 2022. Winter/Spring. 14cm.

'Rainbow Melchior'
Very tall, handsome, early flowering hybrid seedling in the Rainbow 'Three Kings' series. Well-rounded flowers. Outer segments broad, rounded to bluntly pointed apex. Inner segments green inverted 'V'-shaped mark above sinus fading into a vaguely 'X'-shaped mark. Often two scapes to each bulb. Three Kings series, Rainbow Farm. 2020. December, usually around Christmas. 25-30cm.

'Rainbow Selene'
Good green marked *G. elwesii*. Leaves supervolute, broad, erect to arching, grey-green. Outer segments strongly incurved with merging green lines above apex to one third of segment. Inner segments dark green mark above sinus almost to base. Vigorous and bulks up well. Third in 'Green Goddess Series' of green marked *G. elwesii*, and named for the Greek Goddess of the moon. Rainbow Farm. 2022. Winter/Spring.

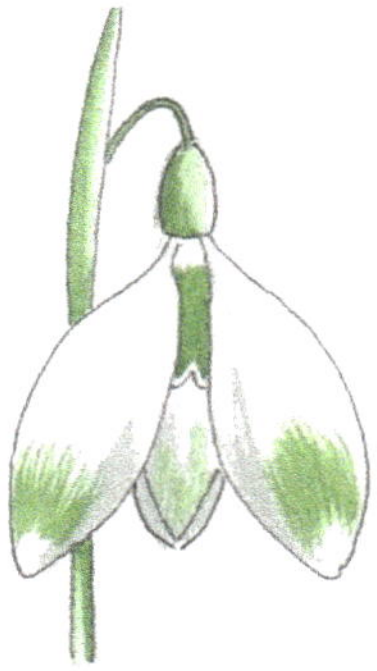

'Rainbow's End'
Good, strong, yellow marked Hybrid with slender pedicel and light-green ovary. Outer segments rounded to bluntly pointed apex. Inner segments yellow inverted 'V'-shaped mark above sinus which can start pale green and get more yellow as flowers mature. Rainbow Farm. Winter/Spring.

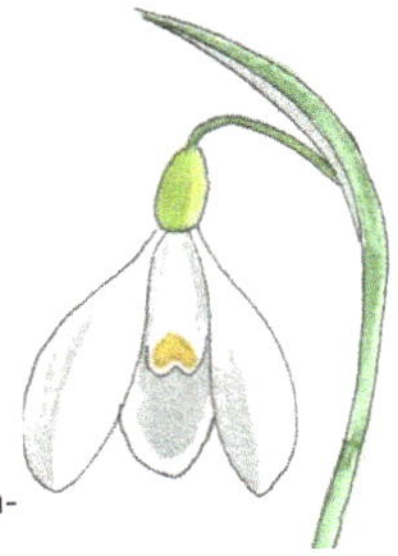

'Rapunzel'
Quirky looking Autumn flowering *G. reginae-olgae* with long, erect scape and small flowers beneath slender, oblong green ovary. Leaves absent at flowering. Outer segments slender, tapering to pointed apex. Inner segments green inverted 'U'-shaped mark to half of segment from above sinus. Easy to see where the name comes from as the slender flower on tall, erect scape is reminiscent of Rapunzel looking out from her high tower. Autumn. 14-16cm.

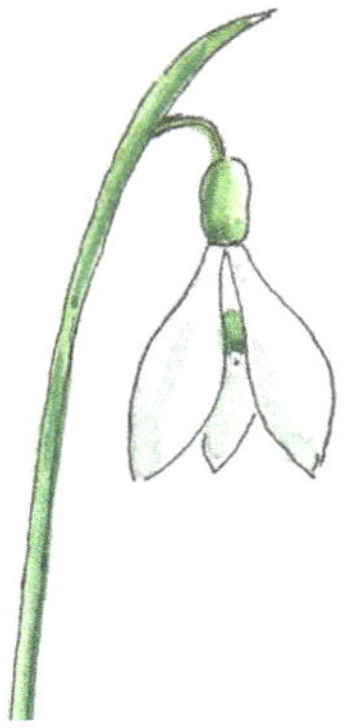

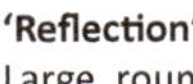

'Reflection'
Large, rounded flowers with oblong green ovary. Outer segments broad, short claw, rounded to bluntly pointed apex, lightly longitudinally ridged. Inner segments green inverted 'U' to 'V'-shaped mark above sinus, paler reflection up towards base.

'Renata Meier'
Attractive green-marked Poculiform snowdrop with rounded green ovary, Segments all of equal length, slender, tapering to bluntly pointed apex, green mark above apex on lower third of segment. Inner segments have a slightly paler mark above apex. Rudi Meier.

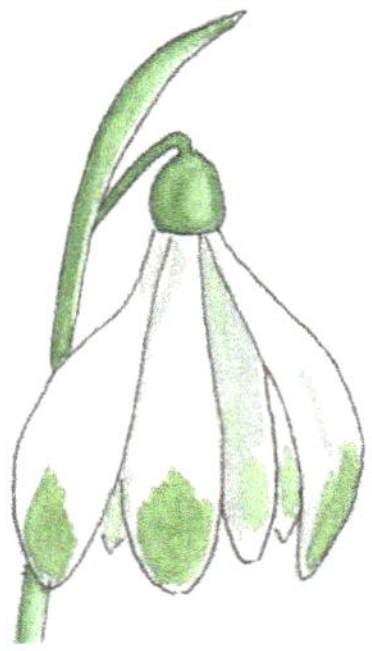

'Richard Nutt's Green-leaved Hybrid''
Very rare, attractive but difficult to obtain. Leaves supervolute and dark green. Outer segments broad, rounded to bluntly pointed apex, longitudinally ridged. Inner segments inverted green 'V'-shaped mark above sinus. Bishop, Davis and Grimshaw conclude it is a hybrid between *G. woronowii* and *G. rizehensis*, introduced by Galanthophile Richard Nutt from Hopa, Turkey. Vigorous. 10-12cm.

'Richard Todd'
Good double snowdrop with rounded flowers and distinctive additional segments protruding from the inner ruff. Leaves erect to arching, convolute, grey-green. Outer segments rounded, tapering to bluntly pointed apex, occasional green marks above apex. Inner segments neat ruff with inverted 'V'-shaped mark above sinus and three fang-like additional segments protruding from the centre. Anglesey Abbey and named for the retired head gardener. Rainbow Farm c. 2012, released 2021.

'Ristpart'
Estonian Bird series of snowdrops. ('Ristpart' = Shell Duck - *Tadorna tadorna*). Long, full, slender, rounded double flowers on short pedicel with rounded ovary. Outer segments slender claw, bluntly pointed at apex. Inner segments loose arrangement with broad, green inverted 'V'-shaped mark above small sinus, second paler mark towards base. Taavi Tuulik, Estonia. Available from Ann Wright, Dryad Nursery. 2020. Winter/Spring. 10cm.

'Robust Timpany' - Syn *G.* 'Robustus'
This Irish hybrid is no longer available under the name *G.* 'Robustus'. Very large plants with large flowers and rounded green ovary. Outer segments rounded, lightly textured. Inner segments broad green heart-shaped mark above sinus. Bulks up well. County Down, Ireland. 18-25cm.

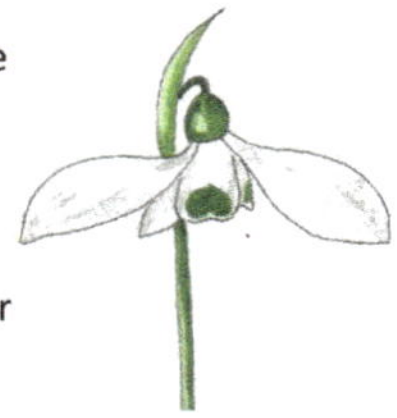

'Rocket Ship'
G. elwesii with very slender, flaring, pointed flowers on erect scapes with small green ovary. Leaves erect, broad, grey-green. Outer segments slender, incurved, pointed at apex. Inner segments green inverted 'V'- shaped mark above sinus, second green mark above almost to base. Cal Mateer, British Columbia. 2019.

'Rockland'
Shorter growing snowdrop with well rounded flowers. Leaves short at flowering, medium, green. Outer segments, short claw, rounded, tapering to pointed apex. Small green mark on either side of apex. Inner segments green mark from above sinus almost to base. Canada. 8cm.

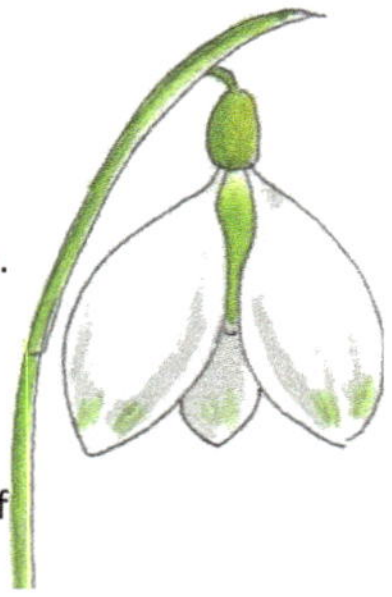

'Rodmarton'
A large and early, well-formed Hybrid cultivar double with rounded flowers of good substance, slightly inflated spathe and long slender ovary. Leaves explicative, semi-erect, two-three, blue-green, paler median line. Outer segments variable, broad, clawed, longitudinally ridged, variable green mark above apex. Inner segment ruff, outer whorl larger, often one aberrant segment, broad green inverted 'U'-shaped mark above sinus, diffusing towards base. One of the earliest flowering double hybrids. *G. plicatus x G. nivalis* 'Flore Pleno'. Bulks up well. Mary Biddulph from her garden, Rodmarten Manor, Gloucestershire. mid-1970s. 26cm.

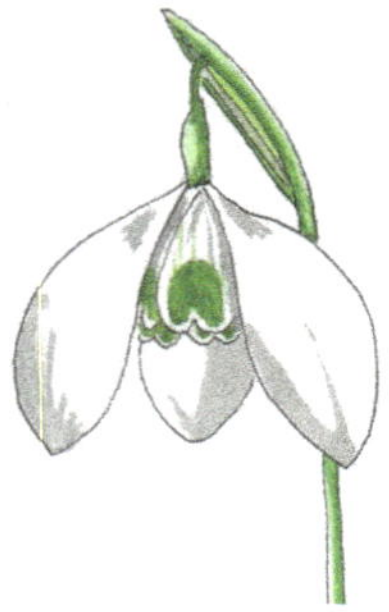

'Roll On Jo'
Heavily green marked flowers on short pedicel between long slender, split spathe and oblong ovary. Outer segments slender, rounded at apex, heavily marked green from apex to base, narrow white margin. Inner segments inverted dark-green 'V'- shaped mark above sinus bleeding towards base, narrow white margin.

'Rose Baron'
Attractive, well-rounded Hybrid cultivar with small green ovary. Leaves erect, blue-green. Outer segments well rounded, rounded at apex, lightly longitudinally ridged. Inner segments broad with rounded green inverted 'U'- shaped mark above sinus and two paler joining ovals towards base. Michael Baron, Brandymount, Hampshire. 15cm.

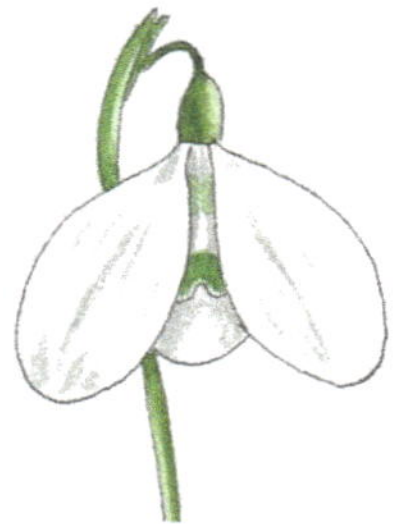

'Rosie'
Classic double flowers. Leaves erect, medium, green. Outer segments rounded, incurved to pointed apex, small pale-green mark above apex. Inner segments neat ruff, broad green inverted 'U'- shaped mark above sinus joining paler mark towards base which can form a roughly 'X'- shaped mark. Hugh and Hilary Purkess's garden, Welshway, Cirencester, Gloucestershire, and named for their granddaughter. 1995. 15cm.

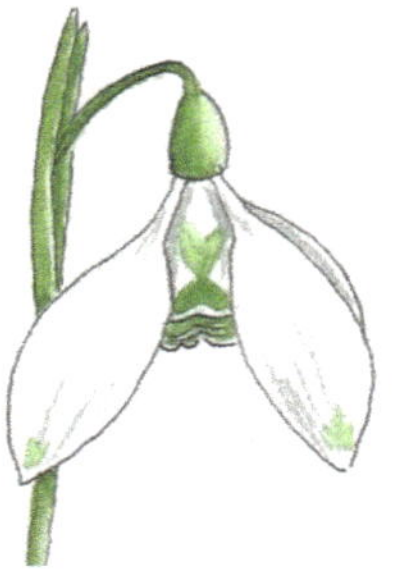

'Roy Buchanan'
Inverse Poculiform snowdrop with slender- looking, green-marked flowers and rounded green ovary. Segments of equal length, slender, rounded to pointed apex, small green lines above apex on lower third of segment.

'Roy's Double'
Unusual double *G. nivalis* with extra segments from centre ruff. Leaves applanate, erect to arching, slender, blue-green. Outer segments flaring, tapering to bluntly pointed apex. Inner segments broad with green marks at sinus, additional white segments protruding from centre of ruff. Similar snowdrop to G. 'Mosquito'. Paul Barney, Edulis Nursery and named for his father. Watlington. 2010. 14cm.

'Ruby's Green Find'
A good virescent *G. nivalis* seedling discovered by the late Ruby Baker. An unnamed snowdrop from her collection, it was passed on to the late Margaret Owen and named by her. Flowers are a typical *G. nivalis* shape, with good green markings. Leaves applanate, erect, slender, blue-green. Outer segments slender, with delicate light-green marks above apex. Inner segments light-green 'W'- shaped mark above sinus with delicate paler lines bleeding upwards. 2002.

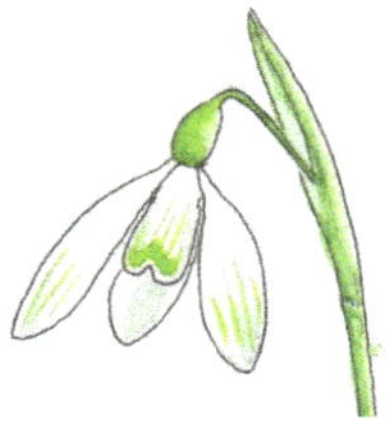

'Rus Ukraine'
Snowdrop with broad spathe, slender pedicel and rounded green ovary. Outer segments rounded, tapering to bluntly pointed apex, small green lines above apex. Inner segments broad green mark above sinus. Ruslan Mishustin, Ukraine.

'Rusian'
Handsome flowers on erect scapes with rounded, oblong, bright-green ovary. Leaves broad, splayed, grey-green. Outer segments broad, short claw, rounded to pointed apex, lightly longitudinally ridged. Inner segments green mark either side of sinus, second mark towards base.

'S F On The Move'
Attractive *G. nivalis* with well-shaped, green marked flowers and oblong ovary. Leaves applanate, erect to arching, slender, blue-green. Outer segments rounded, tapering to pointed apex, short merging green lines above apex. Inner segments inverted 'U'- to 'V'- shaped green mark above small sinus. January.

'Salad Bowl'
G.plicatus subs. *byzantinus* with rounded, lightly textures flowers. Leaves explicative, slender, erect to arching, green, paler median line. Outer segments rounded to bluntly pointed apex, lightly longitudinally ridged and textured. Inner segments, mid-green, slightly waisted mark above sinus almost to base, diffusing on basal edge. Vigorous. Andy Byfield. 2020. Winter/Spring. 16cm.

samothracicus
This new snowdrop was originally discovered in 2006 on the Greek island of Samothraki. Although similar and related to *G. nivalis* there were enough differences for it to be deemed a new species. Leaves erect to arching, slender, blue-green, glaucous, and distinctly greener when mature, losing the glaucous sheen. Flowers are slender and rounded. Outer segments, slender, rounded, slender claw, tapering to bluntly pointed apex, lightly longitudinally ridged. Inner segments broad green inverted 'U'- shaped mark above sinus paling on basal edge. Lightly perfumed. February in cultivation. 16-31cm.

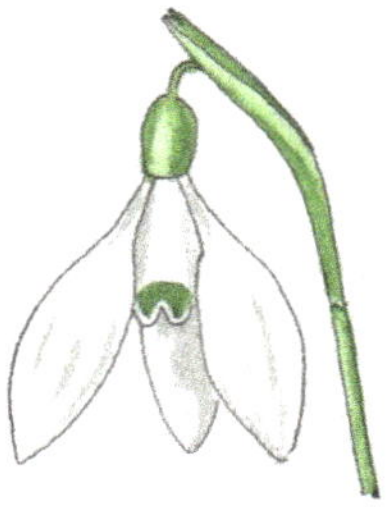

'Sandalio's Charm'
Small, heavily yellow marked flowers on erect scapes with oblong rounded yellow ovary. Segments of equal length, broad, rounded at apex, large lemon-yellow mark above apex. Winter/Spring. 8-10cm.

'Sanger'
Very slender, triangular flowers on erect scapes with slender spathe and long ovary. Inner segments long, slender, incurved, lightly longitudinally ridged and lightly goffered at base, strong green mark above apex to base, shadowed on underside. Inner segments broad, rounded, strong green mark across segment to base, above small sinus, narrow white margin. Germany. 2022.

'SA0901'
Beautiful flowers with delicate green marking on a rather special snowdrop associated with the late David Quinton. Leaves erect to arching, slender, grey-green, paler median line. Outer segments broad, rounded, tapering to bluntly pointed apex, small green lines above apex. Inner segments broad green mark above sinus to half of segment. Winter/Spring. 16-18cm.

'Sarah's Sweet Heart'
Elegant looking *G. elwesii* with large, well-shaped flowers and unusual marking, suspended from short scapes. Leaves supervolute, broad, erect, grey-green. Outer segments tapering to bluntly pointed apex, small green marks above apex. Inner segments narrow green inverted 'V'- to 'W'- shaped mark above sinus joining pale-yellow heart-shaped mark above. Joan Day, USA. 2019. Winter/Spring. 8cm.

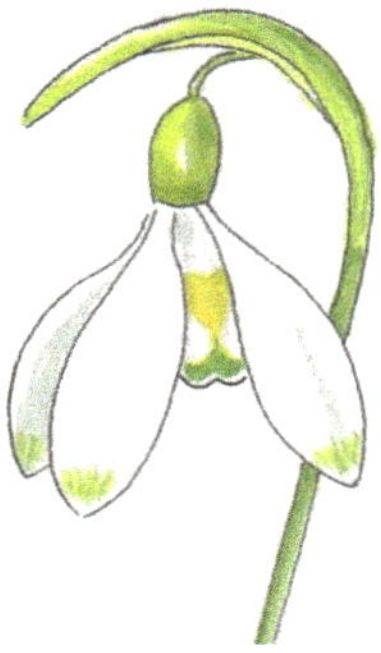

'Scaramouche'
Thought to be a hybrid seedling between *G. woronowii* and *G. elwesii* var. *monostictus* with large, well-shaped flowers. Leaves convolute, matt-green. Well perfumed. Found at Avon Bulbs.

'Scharlocshank'
G. nivalis with split spathe and very long, slender, green-marked flowers with small slender green ovary. Leaves applanate, slender erect to arching, blue-green. Outer segments long, slender with long slender claw, strongly incurved with yellow-green mark above apex to half of segment. Inner segments, broad with wide inverted green 'V'- shaped mark above sinus. Ruben Billiet 2022. 16cm.

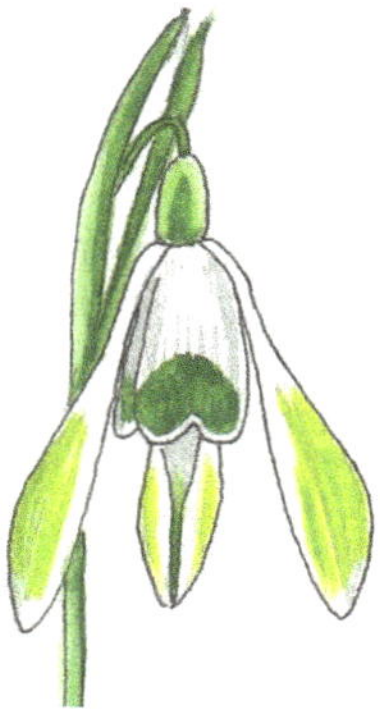

'Schewefelgold'
Fragile looking snowdrop with long, slender, angled pedicel and long, rounded, lemon-yellow, ovary. Leaves narrow, erect to splayed, blue-green, paler median line. Outer segments slender, tapering to pointed apex. inner segments narrow, lemon-yellow mark above sinus. Winter/Spring. 10cm.

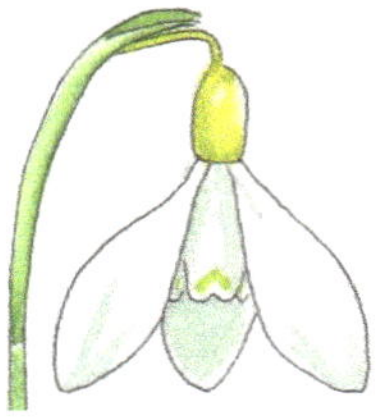

'Schewefeifee' ('Brimstone or Sulphur Fairy')
Slender-looking, yellow marked flowers with long slender yellow ovary. Leaves erect to splayed, slender, green, paler median line. Outer segments tapering to pointed apex. Inner segments brimstone, (yellow-green) - narrow inverted 'U'- shaped mark above sinus. Andreas Handel. Found in woodland near Lingen, by a forester. 2019. Winter/Spring. 12cm.

'Schorbuser Maze' (Wandering Light)
Large, well-shaped, rounded flowers with rounded green ovary contrasting well with strong yellow inner marking. Outer segments broad, well-rounded, short claw, lightly longitudinally ridged, tapering to bluntly pointed apex. Inner segments narrow inverted golden 'V'- shaped mark above sinus and second yellow oval towards base. Winter/Spring. 15cm. Hagen Engelmann, Germany.

'Seratin'
Well-shaped slender looking flowers with oblong green ovary. Outer segments slender, rounded, tapering to pointed apex. Inner segments dark-green, full heart-shaped mark above sinus. Winter/Spring.

'Seremony'
Elegant, green-marked flowers with rounded, oblong green ovary. Could be an Inverse Poculiform or semi-inverse Poculiform according to Joe Sharman. Leaves broad, erect, longitudinally ridged, green. Outer segments flattened, rounded to rounded apex, lightly longitudinally ridged, broad green mark above apex. Possible small sinus. Inner segments shorter than outer with narrow green inverted 'V'- shaped mark above sinus. Joe Sharman. Monksilver Nursery, Cambridge and discovered in the Late Veronica Cross's garden.

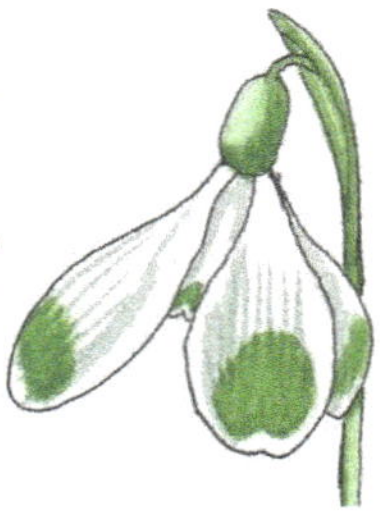

'Shampita'
Larger flowered *G. nivalis* with oblong, dark-green ovary and slender looking flowers. Leaves applanate, erect to arching, slender, blue-green. Outer segments slender, rounded to pointed apex. Inner segments broad dark-green 'W'- shaped mark above sinus.

'Shimmer'
An early *G. elwesii* hybrid seedling from Avon Bulbs which stands out from the crowd with its bright yellow-green colouring. Leaves supervolute, erect, broad, grey-green. Outer segments rounded, tapering to pointed apex, light-green marks above apex on lower third of segment. Inner segments mid-green mark across segment from above sinus almost to base, narrow white margin. Avon bulbs. Autumn/Winter.

'Shrubbery Special'
A rare, vigorous hybrid single snowdrop with large, flaring, slender flowers and rounded green ovary. Outer segments long slender claw, rounded to pointed apex, lightly longitudinally ridged. Inner segments, green mark either side of sinus and second green marks towards base which can merge. Possibly *G. gracilis* and *G elwesii* genes. A seedling selected from Rod and Jane Leeds garden in Suffolk.

'Sibbertoft Soldier'
Stately, well-rounded, lightly textured flowers. Outer segments rounded to bluntly pointed apex, lightly longitudinally ridged and textured. Inner segments green inverted 'V'- shaped mark above sinus, second mark towards base.

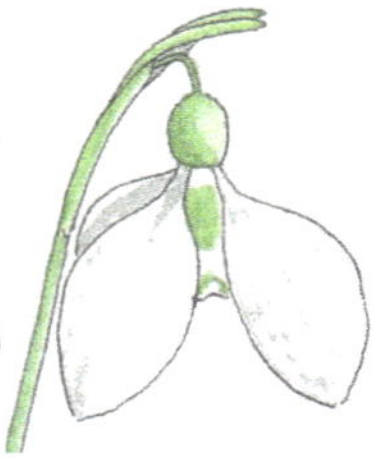

'Silver Star'
Usually a four segmented *G. nivalis*. Leaves applanate, erect to splayed, slender, blue-green, paler median line. Outer segments four, long, slender, to bluntly pointed apex, small green mark above apex. Inner segments roughly inverted 'V' shaped mark above small sinus. 15cm.

'Simone'
G. nivalis with slender oblong green ovary and slender, green-marked flowers. Leaves applanate, erect to splayed, slender, blue-green. Outer segments slender, incurved to pointed apex, light-green stripes above apex to half of segment. Inner segments broad squared green inverted 'U'- shaped mark above sinus. Snowdropfevers. 2021. Winter/Spring. 15cm.

'Sioux'
G. nivalis cultivar with oblong green ovary, throwing additional appendages from the spathe, resembling two Indian head feathers. Leaves applanate, erect to arching, slender, blue-green. Outer segments slender claw, rounded to bluntly pointed apex. Inner segments inverted green 'V'- shaped mark above sinus.

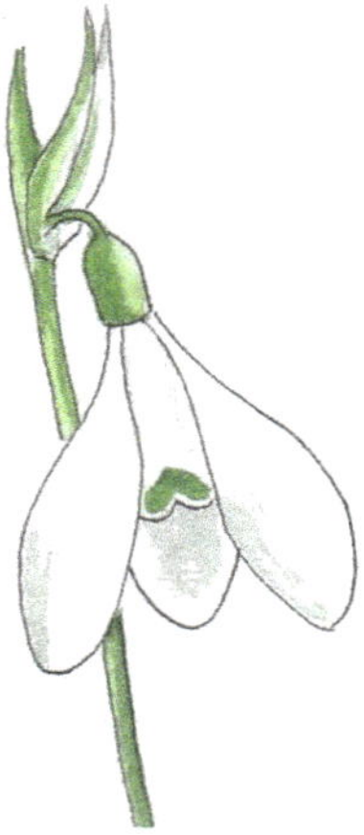

'Sixtus'
G. nivalis snowdrop with, as the name implies, six long slender segments. Leaves applanate, slender, erect to arching, blue-green. Six segments of equal length, slender claw, rounded, tapering to pointed apex, pale green lines above apex on lower third of segment. Winter/Spring.

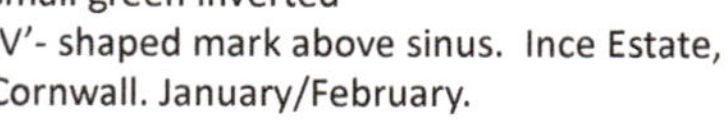

'Snow Angel'
G nivalis Scharlockii with elegant flowers suspended from long slender pedicel between split spathe and green marked segments. Leaves applanate, erect, slender, blue-green. Outer segments rounded to bluntly pointed apex, green mark above apex. Inner segments small green inverted 'V'- shaped mark above sinus. Ince Estate, Cornwall. January/February.

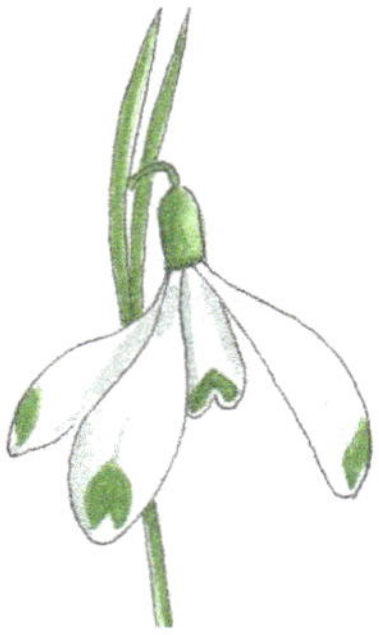

'Snow Flurry'
Elegant snowdrop with very rounded flowers on slender pedicel with rounded, dark-green ovary. Outer segments short claw, rounded, tapering to bluntly pointed apex. Inner segments green mark either side of sinus and second large green mark towards base.

'Snowwhite' ex 'Hulsman'
Large rounded flowers with rounded green ovary. Outer segments short claw, broad, rounded, tapering to bluntly pointed apex. Inner segments narrow green inverted 'V'- shaped mark above sinus, second green oval towards base.

'Solo'
Rounded, pointed flowers suspended from slender pedicel with oblong green ovary. Outer segments broad, clawed, rounded, tapering to pointed apex, green lines above apex on lower third of segment. Inner segments broad inverted 'U'- shaped mark above sinus.

'Spare Bear'
Very elegant looking *G. elwesii* Poculiform with very long, thin buds opening to long, slender, flaring flowers with three segments curving inwards and oblong, dark green ovary. Leaves supervolute, erect, medium, grey-green. Segments slender, of equal length, longitudinally ridged with small green mark above apex. Three segments flaring upwards, three curving inwards. Cal Mateer, British Columbia. Winter/Spring.

'Spili'
Very slender flowers with slender, oblong green ovary. Outer segments long, slender, pointed at apex, lightly longitudinally ridged. Inner segments green mark either side of sinus, second oval at base.

'Spokes Pinwheel'
Very full *G. nivalis* double with flaring segments. Leaves applanate, erect to arching, slender blue-green. Outer segments reflexing upwards, some green tips. Inner segments broad, full ruff with small green inverted 'U'- shaped mark above sinus.Winter/Spring. 10cm.

'Spring Cottage'
Well-textured *G. plicatus* flowers suspended from slender pedicel with rounded green ovary. Leaves explicative, broad, erect to arching, green. Outer segments rounded to bluntly pointed apex, longitudinally ridged and well textured. Inner segments broad green inverted 'U'-shaped mark above sinus merging into two spots towards base. 2017. 16cm.

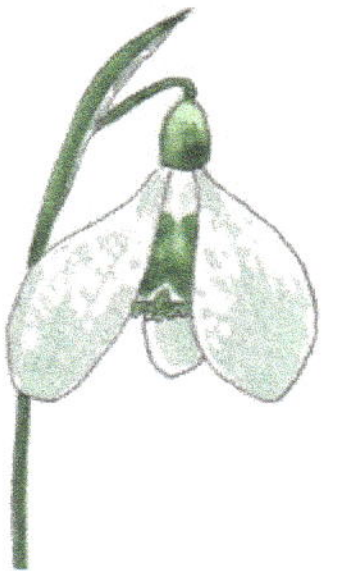

'Sprite'
Attractive virescent with good green marks. Outer segments rounded to bluntly pointed apex, short good green lines in veins above apex. Inner segments broad green mark above sinus to three quarters of segment. Easy and bulks up fast.

'Squidley'
G. nivalis with very slender looking, green marked flowers. Leaves applanate, erect to arching, slender, blue-green. Outer segments long claw, slender, incurved, tapering to pointed apex, green lines above apex. Inner segments broad, rounded, dark-green heart-shaped mark above sinus. Mature flowers said to resemble a squid. Decora Nursery, Croatia.

'St Antonius'
Later-flowering Inverse Poculiform flowers on tall, erect scapes with large, oblong, rounded green ovary. Segments all of equal length, broad, flattened, rounded to apex with inverted mid-green 'V'- shaped mark above sinus and light longitudinal ridging. Germany. March/April. 25cm.

'Starbright'
Inverse Poculiform *G. nivalis* with neat-looking flowers. oblong, yellow-green ovary and yellow-green marks. Leaves applanate, erect to splayed, slender, blue-green, paler median line. Outer segments flattened, tapering to rounded apex, yellow-green mark above slightly pinched apex, lightly longitudinally ridged. Inner segments broad inverse 'U'- shaped mark above sinus. The first inverse 'yellow' Poculiform mentioned in the Daffodil and Tulip yearbook 2011. Jorg Lebsa. Winter/Spring. 16-18cm.

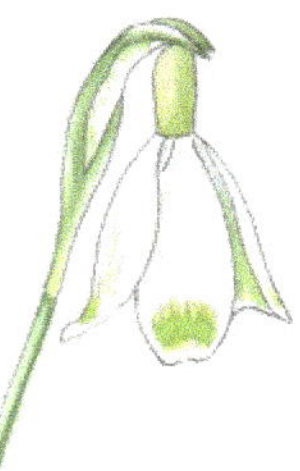

'Steibritz'
See *G.*'Bullfight'.

'Stork'
Lanky-looking *G. nivalis* with slender, outward facing flowers held at an angle with a short pedicel straight from the spathe creating 'stork-like beaks'. Leaves applanate, slender erect, green. Outer segments long, slender, tapering to pointed apex, green mark above apex. Inner segments inverted green 'V'- shaped mark above small sinus. Slow to bulk up. North Green selected seedling from seed originating with a friend in Prague. John Morley, North Green Snowdrops. 2018.

'Strasbourg Twins'
G. x valentinei with two flowers suspended from slender pedicels on one scape. Outer segments rounded to pointed apex, longitudinally ridged. Inner segments green inverted 'V'- shaped mark above sinus and second broad mark towards base. Winter/Spring.

'Streets Ahead'
Handsome large flowered *G. elwesii* seedling with oblong green ovary. Long, rounded outer segments tapering to bluntly pointed apex with small merging green lines above apex. Inner segments, dark green mark above apex almost joining two merging ovals towards base. An Andy Byfield selection from Helen Squires garden, Oxfordshire.

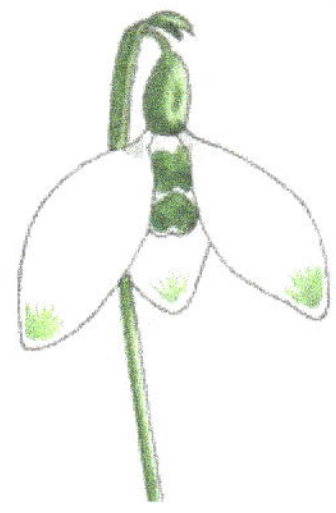

'Streifenzart'
Well-rounded flowers with oblong green ovary suspended from slender pedicel. Outer segments broad, rounded, tapering to rounded apex, green lines above apex to half of segment. Inner segments mid-green mark across segment diffusing slightly at base, narrow white margin.

'Strewelpeter'
Untidy-looking double flowers. Leaves slender, erect to splayed. blue-green. Outer segments slender, curving to pointed apex, variable green marks above apex. Inner segments tight ruff well-marked green. Winter/Spring. 12cm.

'Stuard Turner'
Flowers strongly reflexing upwards. Leaves erect to arching, narrow, green-grey. Outer segments slender, strongly reflexing upwards. Inner segments slender, rounded, narrow green inverted 'V'-shaped mark with rounded ends above sinus. Winter/Spring. 18cm.

'Styx'
Virescent *G. nivalis* with good, green-marked flowers, slightly inflated spathe and oblong green ovary. Leaves applanate, erect to splayed, slender, blue-green, paler median line. Outer segments broad, rounded, tapering to bluntly pointed, pinched apex, green lines above apex to three quarters of segment, light green wash. Inner segments broad, flattened, dark-green mark above sinus to half of segment. Richard Meijndert.

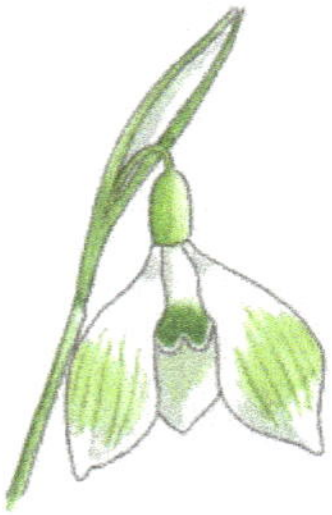

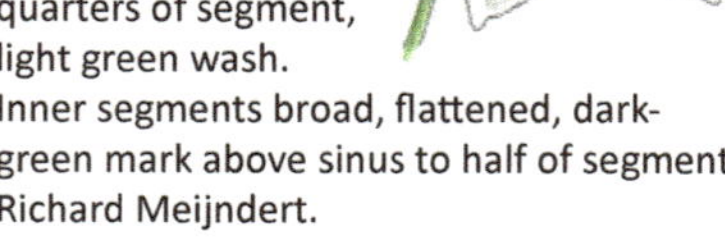

'Sunbeam'
Well-shaped, good strong yellow with yellow-green spathe, yellow pedicel and oblong yellow ovary. Leaves arching-splayed. Outer segments rounded tapering to bluntly pointed apex. Inner segments strong yellow mark above sinus diffusing on basal edge. January-March. 10cm.

'Sunlime'
Good combination of yellow and green with yellow-green spathe, pedicel and ovary. Outer segments rounded, tapering to pointed apex. Inner segments, spreading inverted green 'V'- shaped mark above sinus.

'Sunrise'
G. nivalis with large slender-looking, rounded flowers and oblong green ovary. Leaves applanate, erect to arching, slender, blue-green. Outer segments long, broad, rounded to bluntly pointed apex. Inner segments inverted green 'V'- shaped mark above sinus, diffusing on basal edge. Eddie Tijtgat.

'Svelt'
Elegant mid to late flowering *G. elwesii* with long pedicel, narrow mid-green ovary, and slender flowers. Leaves supervolute, broad, erect, grey-geen. Outer segments slender, rounded, tapering to pointed apex, small green marks above apex. Inner segments broad inverted green 'U'- shaped mark above sinus.

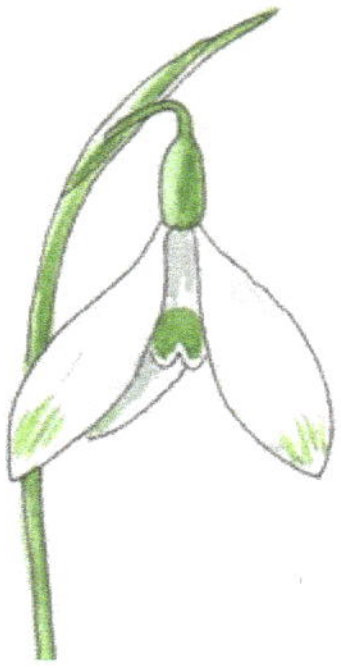

'Sweet Gaia'
G. nivalis with long, green marked flowers and oblong green ovary. Leaves applanate, erect, slender, blue-green. Outer segments long, slender claw, rounded, tapering to bluntly pointed apex, green lines above apex to half of segment. Inner segments broad green mark above sinus diffusing on basal edge. Richard Meijndert and named after his dog.

'Swinging Star'
Perfect *G. nivalis* Poculiform flowers with slender green ovary and exceptionally long, slender, arching pedicel. Leaves applanate, slender, erect to arching, blue-green. Segments all of equal length, rounded to bluntly pointed apex, orange shadow at base. Ruben Billiet and Oliver Vico in an old garden in Northern France, incorporating the 'Star' suffix common to their naming and said to be only the second Poculiform snowdrop with such a long arching pedicel.

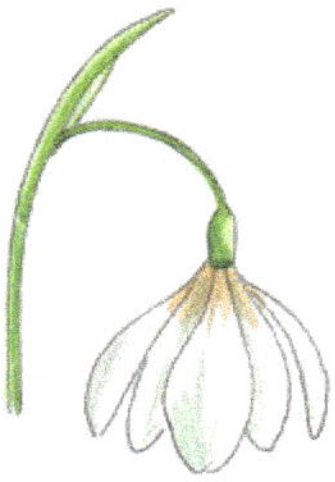

'Sylvan'
Good *G. nivalis* with strong green markings. Leaves applanate, slender, erect to arching, blue-green. Outer segments slender, tapering to bluntly pointed apex, strong green mark above apex. Inner segments green mark across segment from above sinus to base, paling on basal side, narrow white margin. Winter/Spring. 15cm.

'Tangfastic'
A good, late, tall and strong Inverse Poculiform hybrid with rounded green ovary. Segments flattened, rounded at apex, lightly longitudinally ridged with green mark above apex faintly merging into an oval towards base. Paul Barney, Edulis Nursery. February/March.

'Tante Anne'
Sender, pointed, green marked flowers with long, narrow green ovary. Outer segments long slender, incurved, green marks above apex.

'Tatiana Superb'
Weighty-looking, well-rounded flowers on erect scapes with short spathe and slender triangular ovary. Leaves erect, slender, blue-green. Outer segments broad, well rounded, tapering into pinched and pointed apex. Inner segments broad green mark above sinus. Guy de Shriver, Field of Blooms. Winter. 15cm.

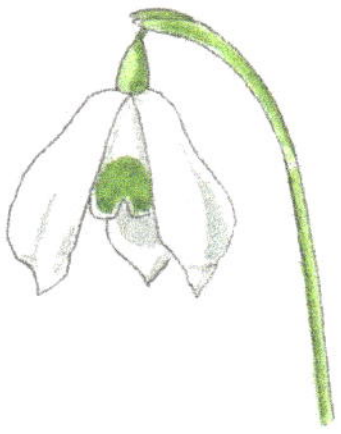

'TCH0110'
Unusual *G. nivalis* snowdrop with broad, split spathe and oblong green ovary. Leaves applanate, erect, slender, blue-green. Outer segments rounded to pointed apex, marked green above apex to half of segment. Inner segments broad green mark above sinus. Ian Christie.

'Ted Kiely'
Well-rounded flowers suspended from slender pedicel with oblong green ovary. Outer segments well rounded, tapering to bluntly pointed apex. Inner segments broad green mark above sinus with second mark above.

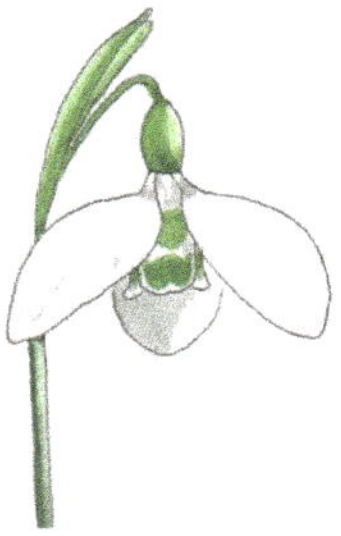

'The Artist'
A later-flowering Inverse Poculiform snowdrop heavily marked green. Long slender pedicel with oblong, lightly ridged, light-green ovary. Segments roughly of equal length with broad light-green mark from above sinus almost to base. 2021. February/March.

'The Green Bucket'
G. nivalis Poculiform snowdrop with well-rounded, green-marked flowers. Leaves applanate, erect. slender, blue-green. Outer segments rounded, tapering to bluntly pointed apex, merging green lines above apex on lower third of segment, diffusing on basal side. Inner segments slightly shorter than outer with merging green lines above sinus. Freddy Van Houtte. Winter/Spring. 14cm.

'The Lady With The Lamp'
G. plicatus with well-rounded flowers and small, rounded green ovary. Leaves explicative, broad, erect to arching, green, paler median line. Outer segments broad, rounded to pointed apex, longitudinally ridged and lightly textured. Inner segments green inverted 'V'-shaped mark above sinus merging into second mark towards base, giving a waisted green mark. Ian Christie and named in honour of Florence Nightingale's 200th anniversary. Money raised from sales of this snowdrop being donated to charity. Winter/Spring 2021. 8cm.

'The Paddle'
Lightly textured, rounded, flowers with rounded green ovary. Outer segments, slender claw, pronounced shoulder, lightly longitudinally ridged and textured, tapering to pointed apex. Inner segments, heart-shaped green mark above sinus, small oval towards base. Winter/Spring.

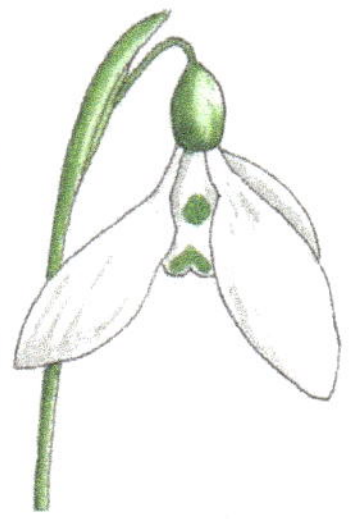

'The Stripper'
Quirky snowdrop with open, spreading flowers of only three outer segments. Leaves erect to splayed. Outer segments, three long, slender, tapering to a point with a pale green stripe along the median line, clearly revealing golden anthers. Gerhard Raschun. 2022.

'Timeless Teresa'
Later, large-flowered hybrid with erect scapes, small, slender, triangular shaped ovary and bold looking flowers which extend the season. Leaves broad, erect, dark-green. Outer segments clawed, long, slender, rounded, tapering to pointed apex, lightly textured. Inner segments broad green inverted heart to 'V'-shaped mark above sinus. Previously sold in error as G. 'Millers Late'. Spring, often well into March.

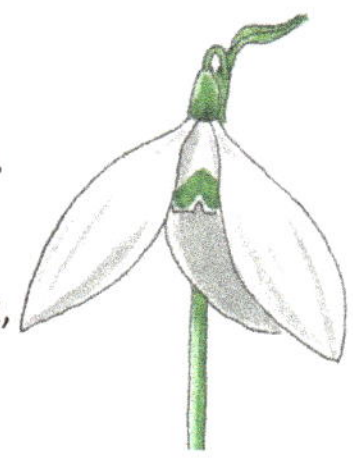

'Timmy Whiteley'
Inverse Poculiform with long pedicel and long, slender mid-green ovary. Segments broad, flattened, rounded at apex, lightly longitudinally ridged, double green mark, narrow crescent above apex and broad green mark towards base. Evenly Wood Gardens. 2021.

'Timpany Bold'
Large-flowered Irish *G. elwesii* var. *monostictus* with attractive foliage. Leaves supervolute, broad, erect to arching, grey. Outer segments clawed, rounded, tapering to bluntly pointed apex. Inner segments bold green mark above sinus. Susan Tindal, County Down. January/February.

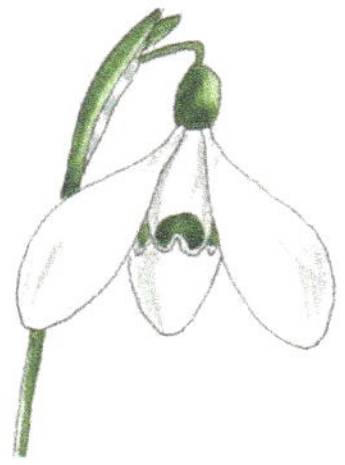

'Tinus Mustache'
A spiky with miscellaneous arrangement of slender, tube-like segments on short pedicel with triangular green ovary. Long slender segment thrown from above ovary, green from apex to base. Outer segments variable, incurved, tube-like, marked green. Inner segments green mark above small sinus. Name originated Tinus-Anna Scheltens, Stone Farm. 12cm.

'Tinus Snor'
Weird-looking *G. nivalis* with split spathe and long, segments protruding from small, green ovary. Leaves applanate, erect to splayed, slender, blue-green. Outer segments long, slender, pointed at apex, heavily marked green with narrow white margin. Inner segments shorter, green inverted 'V'- shaped mark above sinus diffusing towards base. Tinus-Anna Scheltens, Sone Farm. 2019.

'Tiny Sunshine Glow'
G. nivalis with yellow pedicel and ovary and yellow-green spathe. Leaves applanate, slender, erect, blue-green. Outer segments broadening to rounded apex. Inner segments broad yellow mark above sinus. Thorsten Krautwurst. Winter/Spring.

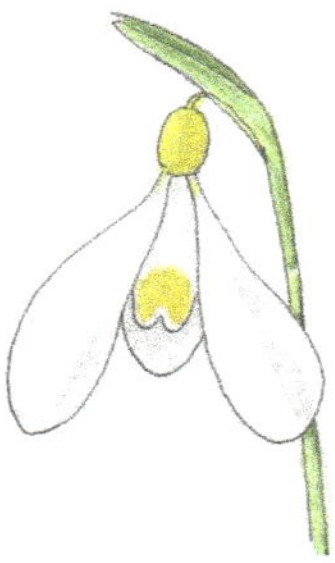

'Tom's Fool'
Unusual looking green-marked *G. nivalis* Poculiform snowdrop with well-rounded, balloon-like flowers and extra segments from oblong green ovary. Leaves applanate, erect, slender, green-blue. Segments broad, rounded to rounded apex, green crescent shaped mark above apex. Winter/Spring. 2021.

'Torgian Geut'
G. plicatus hybrid with inner segment colour changing from green to yellow. Leaves explicative, medium, green-grey. Outer segments broad, rounded, rounded at apex, lightly longitudinally ridged. Inner segments start green and pale to yellow across segment, narrow white margin. Hagen Engelmann 2022. Named for the bells in Torgau, a German town on the banks of the River Elbe, which have rung daily since 1648.

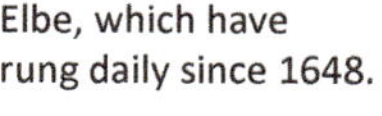

'Tramlines'
G. nivalis with small, oblong, rounded olive-green ovary and spathe. Leaves applanate, erect, slender, blue-green. Outer segments slender claw, rounded to bluntly pointed apex. Inner segments, thin green mark above sinus. Named by North Green Snowdrops. Winter/Spring. 15cm.

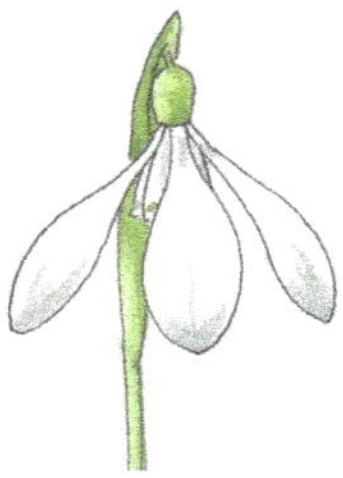

'Trinity'
A stunning *G. plicatus* with large bulbs producing two or three erect scapes. Leaves explicative, medium, green-grey-green. Outer segments broad, clawed, rounded, tapering to bluntly pointed apex, small green lines above apex appearing somewhat like a scallop shell. Inner segments good arching green mark above sinus. Avon Bulbs, 2022.

'Trudy Spotted'
G. nivalis Virescent with good, green-marked flowers. Leaves applanate, erect to splayed, grey-green. Outer segments rounded to bluntly pointed apex, heavily marked green above apex to half of segment. Inner segments green heart-shaped mark above sinus. Discovered by Oliver Vico's wife, Trudy, some years ago. Vigorous. Winter/Spring. 16cm.

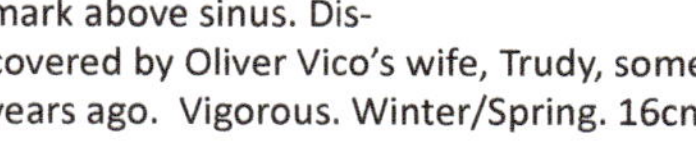

'Truffle'
G. nivalis with neat looking, green-marked, double flowers. Leaves applanate, erect to splayed, slender, blue-green. Outer segments five, flaring, rounded to bluntly pointed apex, variable light-green marks above apex paling towards base. Inner segments, neat ruff with green mark above sinus. Oliver Vico. Winter/Spring. 10-15cm.

'Trumpses'
Slender, green-marked, Inverse Poculiform snowdrop with rounded oblong green ovary. Leaves splayed, slender, blue-green, paler median line. Outer segments slender tapering to rounded apex with sinus notch, green mark above sinus. Inner segments to half of outer with green heart-shaped mark above sinus. Winter/Spring. 15cm.

'Trymskrams'
An Inverse Poculiform with variable green markings. Outer segments broad, flattened, rounded at apex, longitudinally ridged, green mark above apex which can extend to base. Inner segments green mark above sinus which can diffuse towards base. Winter/Spring.

'Tsoe-Tsoe Mama'
Smaller *G. nivalis* spiky double with dumpy, upward facing, green marked flowers on erect scapes with short spathe and small, slender green ovary. Leaves applanate, erect to arching, slender, green-blue, paler median line. Segments variable. Outer segments slightly longer than inner, slender, tapering to rounded apex, marked green above apex towards base. Inner segments loose ruff heavily marked green. Freddy Van Houtte. 2021. Winter/Spring. 8-10cm.

'Tuff'
Rounded flowers on erect scapes with slender pedicel and unusual green-brown coloured ovary. Leaves erect, slender to splayed, green. Outer segments clawed, rounded to bluntly pointed apex. Inner segments narrow inverted green-brown 'V'- shaped mark above sinus. Bulks up well and quickly. 16cm.

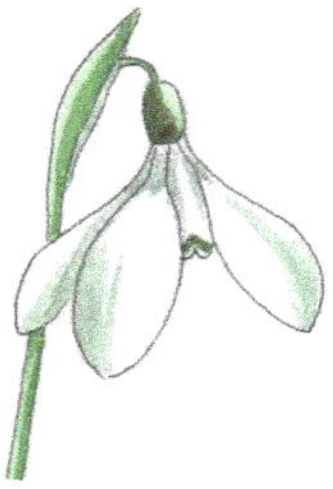

'Tumbledown'
Large and attractive, Virescent G. nivalis snowdrop, well marked green suspended from long slender pedicel. Leaves applanate, slender, erect to arching, blue-green. Outer segments slender, tapering to bluntly pointed apex, green lines above apex to three quarters of segment. Inner segments green mark across segment from above sinus almost to base, narrow white margin. Seedling from G. 'Nova Gorica'. Vigorous, bulks up well. 18cm.

'Tuppence'
Small, neat double *G. woronowii* on erect scapes. Leaves supervolute, broad, arching, green. Outer segments rounded to bluntly pointed apex. Inner segments good green mark above sinus. Originated in bulbs purchased at auction by Gill Gregory and said to be named for its price. 8cm.

'Tutkas'
Estonian bird series of snowdrops - 'Tutkas' - Ruff - *Philomaxchus pugnax* - Slightly inflated spathe with short pedicel and triangular, rounded green ovary. Outer segments, reflexed, slender, rounded and bluntly pointed at apex. Inner segments neat rough with inverted green 'V'- shaped mark above sinus merging into broad green mark diffusing towards base. Taavi Tuulik, Estonia. 2020.

'Tuuliku 8'
Attractive, smaller-growing yellow marked snowdrop with yellow ovary and yellow-tinged pedicel and spathe. Leaves erect, slender, green-blue, paler median line. Outer segments short claw, tapering to pointed apex. Inner segments pale-green mark above sinus becoming yellow and shading towards base. Originally from Taavi Tuulik, via Sulev Savissaar. c.2011. Winter/Spring. 8-10cm.

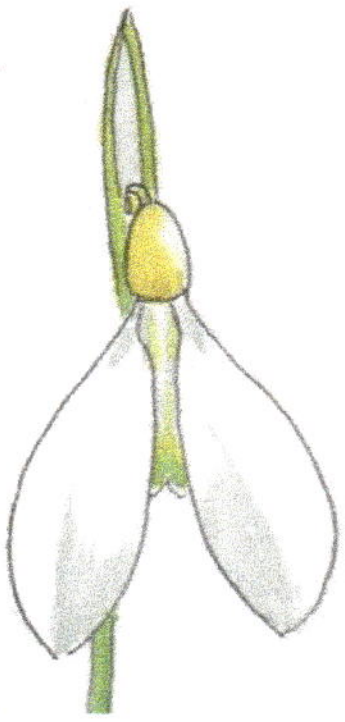

'Twilight'
Shapely flaring flowers with slender segments, good green marks and yellow-green pedicel and ovary. Leaves broad, arching green-grey. Outer segments slender, flaring tapering to pointed apex, lightly longitudinally ridged. Inner segments narrow green inverted 'V'- shaped mark above sinus, second yellow-green mark towards base diffusing on basal side. Richard Bashford. Woodchippings. Winter/Spring.

'Two Tone'
Slender yellow-green ovary and pedicel. Leaves slender, erect, grey-green. Outer segments spoon-shaped to bluntly pointed apex. Inner segments dark-green inverted 'U'- shaped mark above sinus. Snowdropfevers. Winter/Spring. 15cm.

'Uli's GIft'
Colour-change snowdrop, short in stature with large but interesting gold-tinged flowers and good yellow marking. Leaves medium, erect to arching, green. Outer segments broad, rounded to bluntly pointed apex, strongly longitudinally ridged, flushed yellow as flowers mature. Inner segments narrow, spreading, green inverted 'U'- shaped mark with rounded ends above sinus and second oval towards base, both marks change to gold/mustard colour as plants mature. Grows well. 2022. Winter/Spring.

'Uranium'
A *G. reginae-olgae*, said to be the first snowdrop of its type with yellow-green flushed segments with a distinctive green glow. It is similar to *G.* 'Adamite' but the colouring lasts longer. Outer segments rounded to bluntly pointed apex, flushed pale yellow-green across segment. Inner segments green mark above sinus diffusing on basal edge. Joe Sharman, Monksilver Nursery. Cambridge. 2018. Autumn.

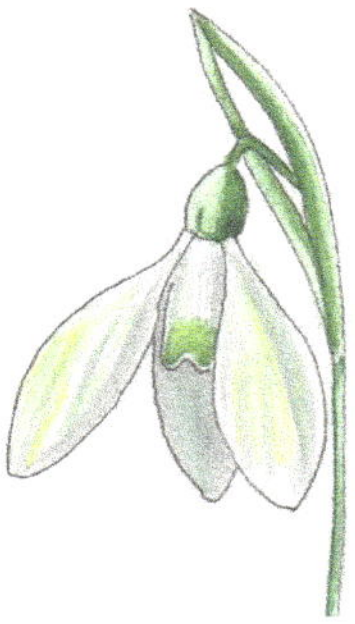

'Uwe Wagner'
G. nivalis with short scapes and heavily green marked flowers suspended from long pedicel between broad, split spathe. Leaves applanate, narrow, splayed, blue-green. Outer segments green wash from apex to base. Inner segments solid green mark from above sinus to base, narrow white margin. 2021. Winter/Spring.10cm.

'Vart'
Estonian Bird series of snowdrops. 'Vart' - Greater Scaup - *Aytha marila*. Triangular ovary with flaring, reflexing outer segments, rounded at apex. Inner segments variable marks from broad inverted 'U'- shaped mark above sinus to full mark across segment. Taavi Tuulik, Estonia. 2020.

'Veliki Progasti'
G. nivalis with large flowers on short scapes. Leaves erect, slender, green. Outer segments long, tapering to bluntly pointed apex, short green lines above apex. Inner segments green horseshoe shaped mark above sinus. Ljubljana Botanic Garden, Slovenia. Winter/Spring. 8cm.

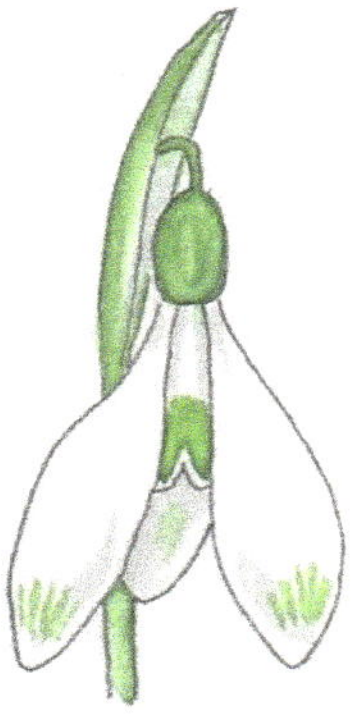

'Vlaam Re'
Shorter-growing snowdrop with very long, slender flowers, oblong green ovary and said to have the longest green-tipped outers. Leaves erect, slender, green-blue, paler median line. Outer segments very long, slender, tapering to pointed apex, lightly longitudinally ridged, yellow-green marks above apex. Inner segments broad green inverted 'V'- shaped mark above sinus, diffusing slightly on basal edge. 2021. Winter/Spring. 8-10cm.

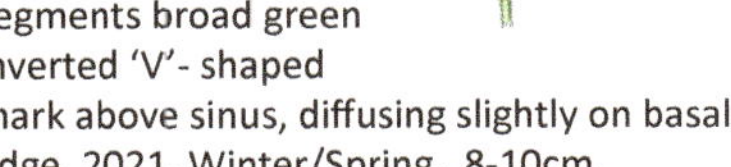

'Walkden Spiky'
G. nivalis spiky with numerous thin, green marked segments.

'Walnut'
Large rounded flowers on erect scapes with slender curving pedicel. Outer segments broad, rounded, rounded at apex, lightly longitudinally ridged and textured. Inner segments green mark from apex almost to base, narrow white margin. A cross between *G.* 'Diggory' and *G.* 'South Hayes' and named as the rounded, textured segments suggest a walnut. Germany c. 2013. 12cm.

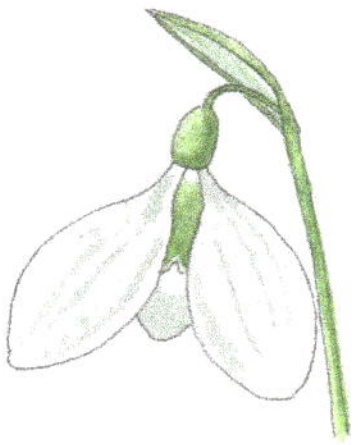

'Walter's Double'
G x valentinei with small, rounded, upward facing double flowers held on short spathe with small green ovary. Leaves applanate, erect, slender, glaucous, with strong paler median line. Outer segments short, rounded to rounded apex, short green lines above apex. Inner segments a neat rectangular rosette, green mark across segment from above sinus slightly diffusing at base, narrow white margin. Walter and Celia Saunders. 2021. 10cm.

'Waltham Place 2'
G. nivalis with slender-looking flowers and oblong green ovary. Leaves applanate, erect, slender, blue-grey. Outer segments slender, rounded, bluntly pointed at apex, inner segments broad inverted green 'V'- shaped mark above sinus merging into an oval towards base. February. 12cm.

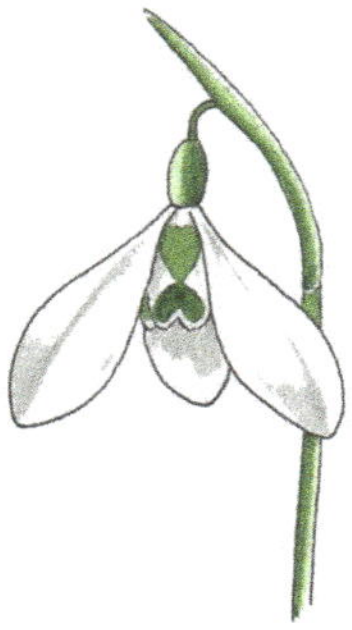

'Wantage Green'
G. plicatus with very green, broad, erect, leaves with, paler median line. Outer segments rounded to bluntly pointed apex. Inner segments green mark above sinus. Very vigorous, bulks up well and excellent for naturalising. Winter/ Spring. 18-20cm.

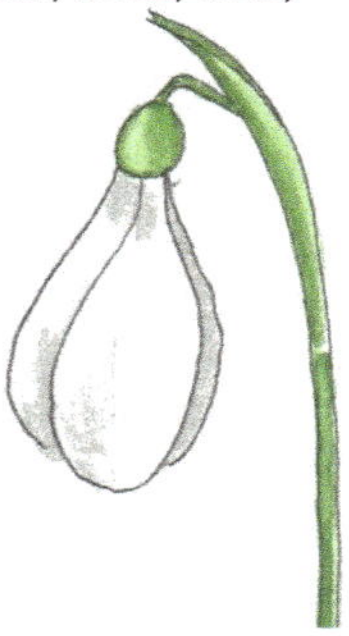

'Warwickshire Gemini'
Good, tall and vigorous *G. elwesii* often having two flowers to a scape when established. Leaves supervolute, broad, erect to arching, grey-green. Outer segments long, spoon-shaped, bluntly pointed at apex. Inner segments green inverted 'U'- shaped mark above sinus. 25cm.

''Wassergeist' - ('Water Ghost')
Unusual, quirky-looking flowers throwing additional broad segments from the ovary. Leaves slender, erect to arching, blue-green. Long, slender buds opening into slender rounded flowers with slender, rounded, incurved outer segments. Germany. Winter/Spring. 10cm.

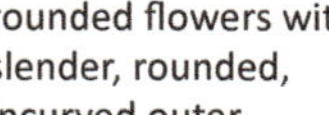

'Watersmeet'
Smaller *G. nivalis* with upward facing flowers and small, slender green ovary. Leaves applanate, erect, slender, blue-green. Outer segments rounded, tapering to rounded apex. Inner segments narrow green inverted 'V'- shaped mark above sinus. Named as it was found where the North Tyne meets the South Tyne rivers. 8-10cm.

'Watlington Greenman'
Large and attractive *G. elwesii* with very erect scapes. Leaves supervolute, erect, broad, grey-green. Outer segments slender tapering to pointed apex. Inner segments, green inverted 'V'- shaped mark above sinus joining an elongated mark almost to base, covering most of segment. Vigorous and bulks up well. Paul Barney, Edulis Nursery. 2019. 18cm.

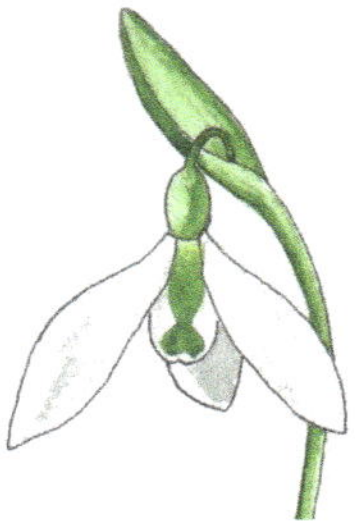

'Watlington Smudge'
Large *G nivalis* with rounded, strongly green-marked flowers. Leaves applanate, slender, erect to splayed, blue-green. Outer segments rounded to rounded apex, longitudinally ridged, merging green lines above apex with faint yellow smudging. Inner segments lightly longitudinally ridged, mid-green 'W'- shaped mark above sinus diffusing lightly on basal edge and running into ridges. Paul Barney, Edulis Nursery. 2020. Winter/Spring. 16cm.

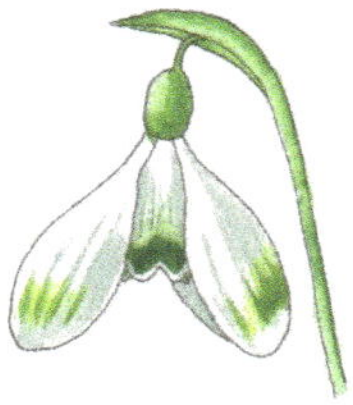

'Watson L'
G. elwesii with rounded, balloon-like, green-marked flowers and oblong green ovary. Leaves supervolute, erect, broad, grey-green. Outer segments, broad, rounded, incurved, tapering to pointed apex, lightly longitudinally ridged, short merging green lines above apex, Inner segments green mark above sinus. Winter/ Spring.

'Well Hung'
Very slender pointed flowers on long, slender, arching pedicel with narrow oblong ovary. Outer segments long, slender tapering to pointed apex. Inner segments green mark above sinus.

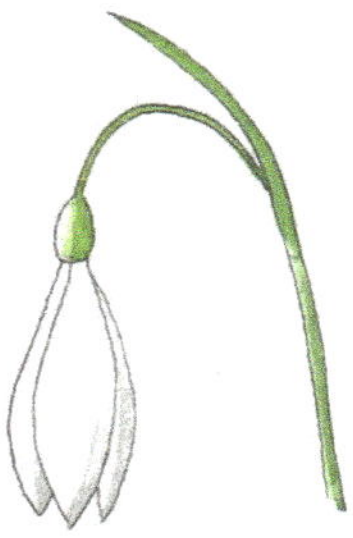

'Wessex Advent'
Early flowering, bold, *G. elwesii*. Leaves supervolute, broad, grey-green. hooded. Outer segments large, incurved, rounded to bluntly pointed apex. Inner segments green mark above sinus and second mark towards base. Paul Barney, Edulis Nursery. December/January. 15cm.

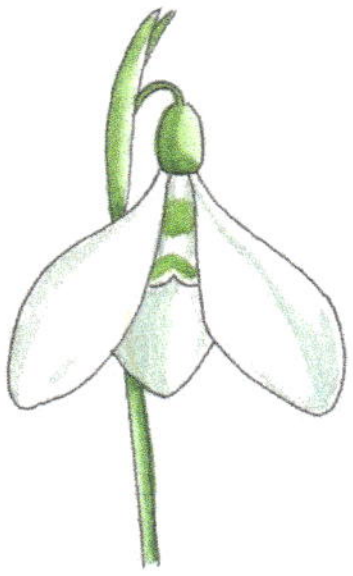

'Wessex Head'
A tall, strong, robust *G. x valentinei* with rounded green ovary and well-shaped flowers. Outer segments broad, clawed, pronounced shoulder, lightly textured, rounded to bluntly pointed apex. Inner segments broad with dark-green inverted 'U'- shaped mark above sinus. Named as it stands 'head and shoulders above the rest'. Paul Barney, Edulis

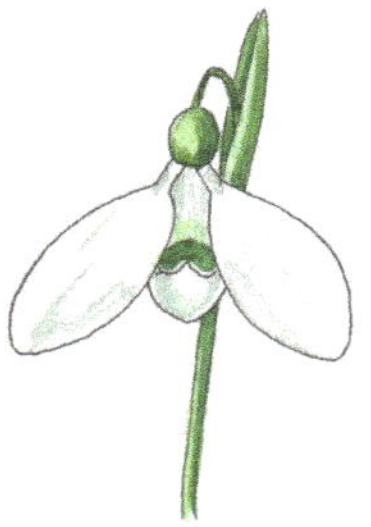

'WeWei Caribou'
Small, rounded *G. elwesii* Poculiform flowers with rounded green ovary. Leaves supervolute, erect, broad, grey-green. Outer segments broad, rounded, incurved to rounded apex, longitudinally ridged, Green mark above apex. Inner segments smaller than outer green mark above apex. Cal Mateer, British Columbia. 2022.

'Whirlng Dervish'
This certainly is, as the name implies 'whirling' with segments flying out in all directions, slender pedicel and oblong green ovary. Outer segments slender, reflexing, incurved, tapering to pointed apex with small green marks above apex. Inner segments variable with green mark above apex or small sinus. Bulks up well.

'White Gold'
Attractive *G elwesii* var. *monostictus* with rounded flowers on slender arching pedicel. Outer segments short claw, rounded to bluntly pointed apex, incurved. Inner segments small yellow-green mark either side of sinus. Freddy Van Houtte. Autumn/Winter. 16cm.

'Whiteheart'
G nivalis with rounded, oblong green ovary and green marked flowers. Leaves applanate erect, slender, blue-green. Outer segments broad, rounded, tapering to bluntly pointed apex, lightly longitudinally ridged, merging green lines above apex on lower third of segment. Inner segments green Chinese-bridge shaped mark above sinus, two small 'eye' spots towards base. Angela Petrisevac. Winter/Spring.

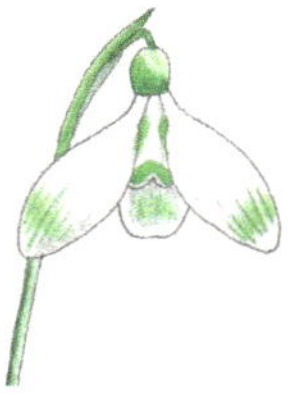

'White Walker'
G. nivalis with erect scapes and long slender flowers. Leaves applanate, slender, splayed, blue-green. Outer segments slender, tapering to bluntly pointed apex. Inner segments narrow, green Inverted 'V'-shaped mark above sinus. Ruben Billiet. Winter/Spring. 15cm.

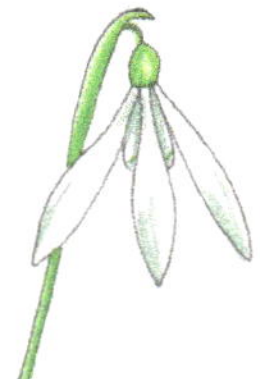

'Widder'
Looks like an ordinary *G. nivalis* Scharlockii of the Ardennes type. To begin with the somewhat leafy spathe valves are straight but as the flower matures the spathe valves curl, resembling a ram's horns, although it is different with each flower. Outer segments slender, tapering to pointed apex. Inner segments narrow inverted green 'V'- shaped mark above sinus. 'Widder', German for ram. Jan Pleuger.
Winter/Spring.

'WiFi Aquarel'
Later *G. nivalis* with delicately marked, well-shaped flowers and small oblong, rounded ovary. Leaves applanate, erect, slender, blue-green. Outer segments long, incurved to bluntly pointed apex, Green lines on lower third of segment with light green wash. Inner segments green mark above sinus almost to base, narrow white margin and distinctive delicate dark-green stripes. Coolplants, Belgium. 2022. March/April.

'WiFi Arial'
Large-flowered *G. nivalis* with green-marked flowers and oblong green ovary. Leaves applanate, medium, grey-green. Outer segments rounded and slightly pinched at apex, light-green mark above apex. Inner segments narrow inverted 'V'- shaped mark above sinus. Named for Arial letter font. Coolplants, Belgium.

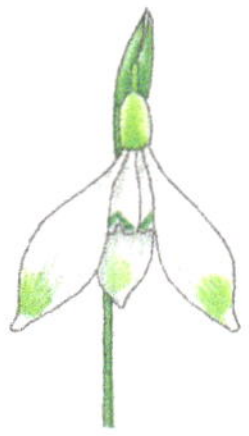

'WiFi Bahnschrift'
Later-flowering *G. nivalis* with slender flowers and long slender green ovary. Leaves applanate, erect, slender blue-green. Outer segments slender, rounded to bluntly pointed apex, green lines above apex. Inner segments inverted green 'U'- shaped mark above sinus, two paler ovals towards base. Named for the Barnschrift letter font. Coolplants, Belgium. 2021. March/April. 15cm.

'WiFi Benji'
G. nivalis similar in appearance to 'WiFi Aquarel' but with flowers suspended from a long, angled pedicel. Leaves, applanate, tall, slender, blue-green. Outer segments incurved, bluntly pointed at apex, green lines above apex to half of segment, light green wash. Inner segments darker green inverted 'V'- shaped mark above sinus with darker edges running towards base and distinctive, strong green stripes. Coolplants, Belgium. 2022. March/April. 16cm.

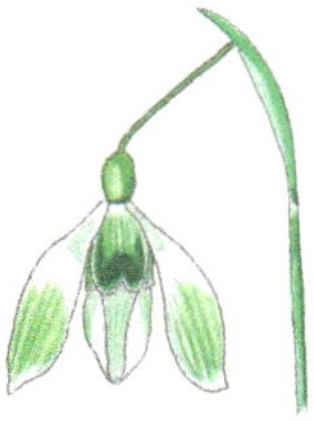

'WiFi Big Brother'
Large *G. elwesii* with weighty-looking flowers suspended from slender pedicel and oblong green ovary. Leaves supervolute, broad, erect, blue-green. Outer segments broad, short claw, rounded tapering to pointed apex, longitudinally ridged with merging green lines above apex. Inner segments broad, rounded with joining green marks above small sinus to three quarters of segment, forming a chalice shaped mark. Named as it is a similar looking snowdrop to *G.* 'Big Boy'. Coolplants, Belgium. January. 18cm.

'WiFi Bingo'
Yellow tinged flowers with oblong yellow-green ovary. Outer segments rounded, tapering to pointed apex. Inner segments good yellow mark above sinus. Seedling from *G.* 'Blonde Inge'. Coolplants, Belgium.

'WiFi Borderline'
Good clean-looking virescent flowers beneath oblong green ovary. Leaves erect to arching, slender, blue-green, paler median line. Outer segments rounded to bluntly pointed apex, good clear green lnes along ridges leaving white border. Inner segments green mark across segment from above sinus diffusing at base. Coolplants, Belgium. 15-20cm.

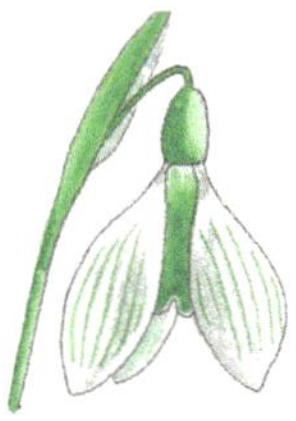

'WiFi Calibri'
Later-flowering *G. nivalis* with long green-marked segments. Leaves applanate, erect to splayed, blue-green. Outer segments long, curving into bluntly-pointed apex, longitudinally ridged, merging green lines above apex on lower third. Inner segments broad, inverted green 'V'- shaped mark above sinus. Named for the Calibri letter font. Coolplants, Belgium. 2021. March-April. 16cm.

'WiFi Caret'
Attractive later-flowering hybrid similar to *G.* 'WiFi Dotty', but later flowering and with differently shaped inner mark. Elegant flowers suspended from long slender pedicel and oblong slender rounded green ovary. Leaves medium, erect to arching, green. Outer segments tapering to bluntly pointed apex, lightly longitudinally ridged and textured. Inner segments, broad, small curving green mark above sinus with incurved ends, paler green mark towards base. February-March. Coolplants, Belgium. March.

'WiFi Century'
Large, rounded flowers with good green marks. Outer segments broad, rounded to bluntly pointed apex, green lines above apex to half of segment. Inner segments inverted green 'U'- shaped mark above sinus. Coolplants, Belgium. 2021.

'WiFi Comic Sans'
Slender-looking flowers with oblong ovary. Leaves slender, erect, blue-green. Outer segments slender, rounded towards bluntly pointed apex, green lines above apex to half of segment. Inner segments broad inverted 'U'- shaped mark above sinus merging into paler lines towards base. Coolplants, Belgium. 2021. 15-18cm.

'WiFi Dipping'
Good virescent *G. nivalis*. Leaves applanate, erect, slender, blue-green with paler median line. Outer segments incurved, rounded to bluntly pointed apex, light green wash across segment to base, above pronounced white apex. Inner segments darker green mark above sinus to base, narrow white margin. Coolplants, Belgium. Winter/Spring. 15-16cm.

'WiFi Dotty'
Included in Volume One as *G.*'Dotty' but included again here with the WiFi prefix for clarity of identification. Attractive hybrid with well-shaped flowers with oblong mid-green ovary. Outer segments, slender, long, rounded, lightly longitudinally ridged, tapering to bluntly pointed apex. Inner segments to half of outer, lightly longitudinally ridged with variable, small, mid-green dot either side of sinus which can merge into a spreading inverted 'U'- shaped mark. Clumps up well. Belgium. February.

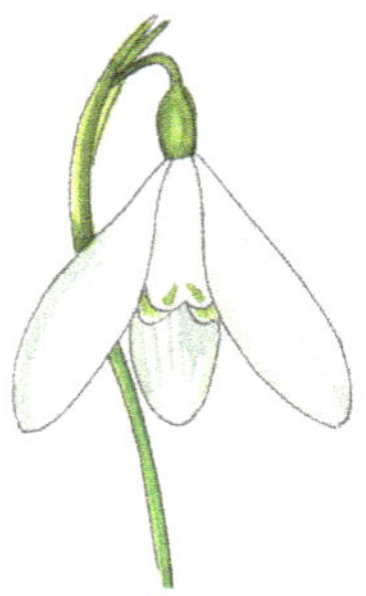

'WiFi Eden's Basket'
Strongly incurved segments resembling a basket shape, hence name, with long slender green ovary. Outer segments strongly incurved, bluntly pointed at apex. Inner segments narrow green inverted 'V'- shaped mark above sinus. Coolplants, Belgium.

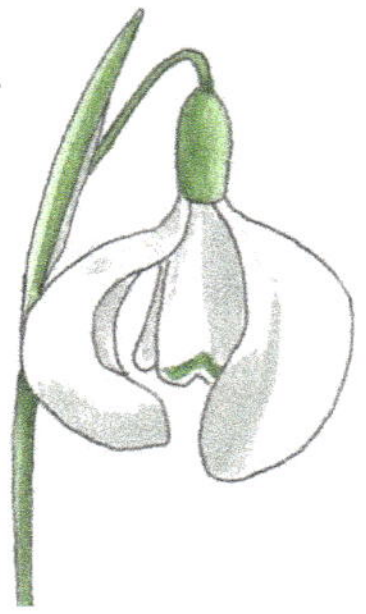

'WiFi Flair'
Later *G. nivalis* with good, large, bell-shaped green-marked flowers helping extend the season. Leaves applanate, erect to arching, blue-green, paler median line. Outer segments long, tapering to pinched apex, green lines above apex on lower third, light-green wash. Inner segments inverted dark-green 'V'- shaped mark above sinus. Coolplants, Belgium. 2022. March/ April. 16-18cm.

'WiFi Flight of Hearts'
Beautiful *G. elwesii* var. *monostictus* with perfect heart-shaped marking. Leaves supervolute, broad, erect, grey-green. Outer segments flaring, bluntly pointed at apex, variable small green marks above apex. Inner segments broad, rounded, with good heart-shaped green mark above sinus. Coolplants, Belgium. Autumn/Winter.

'WiFi Green Bride'
Later *G. nivalis* with large, shapely flowers well marked with green. Leaves applanate, slender, arching, blue-green. Outer segments long, tapering to pinched apex, merging green lines above apex. Inner segments broad inverted dark-green 'U'- shaped mark above sinus. Said to be named as it resembles an A-line wedding dress in shape. Coolplants, Belgium. March/ April. 18cm.

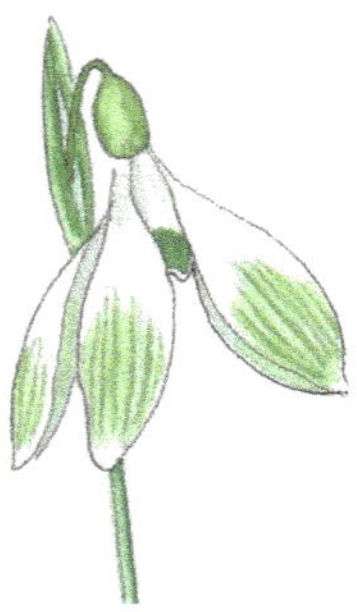

'WiFi Kiruha'
Smaller *G. nivalis* with broad, erect, grey-green leaves and heavily green marked flowers. Leaves applanate, erect, slender, blue-green. Outer segments incurved, bluntly pointed at apex, washed green across segment. Inner segments green mark above sinus to base, narrow white margin. Coolplants, Belgium. 2022. 10cm.

'WiFi Kittiwake'
G. elwesii with strong green marking. Slender, oblong dark-green ovary and long, slender arching pedicel. Leaves erect to arching, broad, grey-green. Outer segments broad, short claw, tapering to pointed apex, lightly longitudinally ridged with strong green mark above apex to half of segment. Inner segments broad green mark across segment from above sinus almost to base, narrow white margin. Coolplants, Belgium. 2022. Winter/Spring.

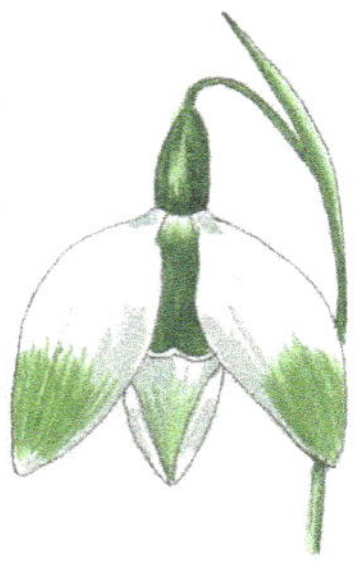

'WiFi Let's Dance'
G. nivalis with large spathe and good, green-marked flowers suspended from long pedicel. Leaves applanate, medium, arching, grey-green. Outer segments, short claw, rounded to bluntly pointed apex, green lines above apex towards base, light green wash. Inner segments broad green mark above large sinus, diffusing towards base. Coolplants, Belgium 2022. February/March. 16cm.

'WiFi Lucida'
Later-flowering snowdrop. Leaves slender, green-grey. Outer segments spoon-shaped with merging green lines above apex to half of segment. Inner segments narrow inverted green "V'-shaped mark above sinus. Named for the Lucida letter font style. Coolplants, Belgium. 2021. March/April. 15cm.

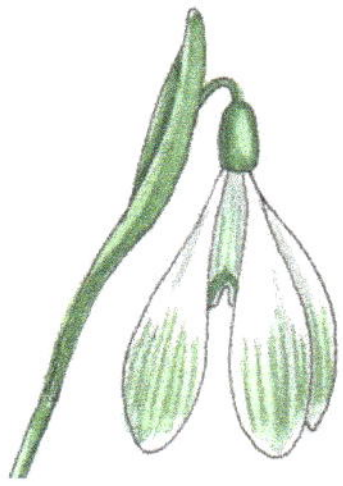

'WiFi Magic Mint'
Attractive *G. nivalis* with green-marked flowers. Leaves applanate, erect to splayed, slender, blue-green. Outer segments slender claw, rounded to bluntly pointed apex, mint green lines almost to base from above white apex. Inner segments mid-green mark across segment above sinus diffusing to base, narrow white margin. Coolplants, Belgium. 16-18cm.

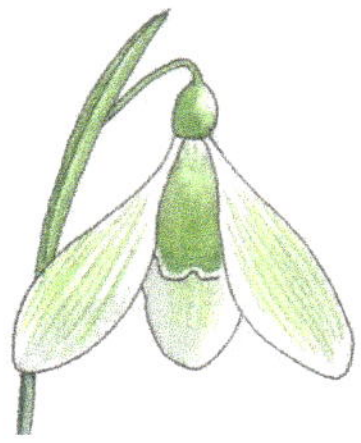

'WiFi Monroe'
A *G.plicatus* with pagoda shaped flowers. Leaves erect to arching, explicative, green-grey, paler median line. Outer segments broad, flaring upwards resembling Marylin Monroe's dress, hence name, tapering to rounded, upward curving apex, small dark-green mark above apex. Inner segments dark-green inverted 'V'- shaped mark above sinus merging into two slightly paler arms, Coolplants, Belgium 2022. February. 15-20cm.

'WiFi Myriad'
Hybrid Virescent snowdrop with slender pedicel and oblong green ovary. Outer segments rounded, tapering to pointed apex, light-green lines across two thirds of segment with paler wash. Inner segments green mark above sinus. Coolplants, Belgium

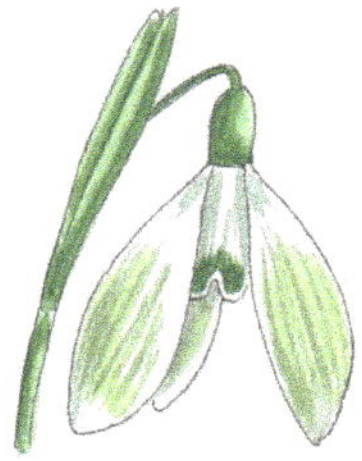

'WiFi Nacitti'
Attractively shaped, well-marked flowers with broad, erect green leaves. Outer segments broad, rounded, short claw, pinched and pointed at apex, green mark above apex. Inner segments full green mark apex almost to base, narrow white margin. Coolplants, Belgium. 2021.

'WiFi Nigeria'
Later flowering *G. nivalis* with good green marks. Leaves applanate, slender, erect, green-blue, paler median line. Outer segments long, rounded to bluntly pointed apex. Five-six green lines above apex on lower third of segment. Inner segments green mark from above sinus diffusing to base. Named for the Nigerian flag. Bulks up well. Coolplants, Belgium. March/April. 16cm.

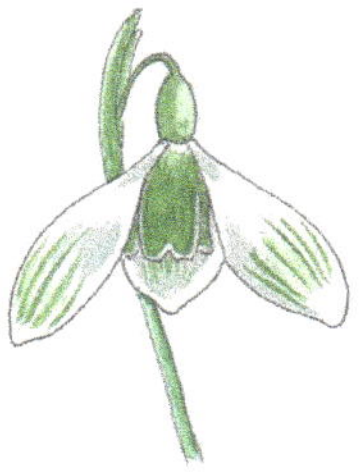

'WiFi Overflow'
Later, large *G. nivalis* with long shapely flowers suspended from long slender pedicel and good green marks that flow upwards from apex of outer segment, hence name. Outer segments broad, long, rounded and slightly flaring at apex with green lines from apex to half of segment, shadowed on underside. Inner segments narrow inverted 'V'- shaped mark with squared ends above sinus. Coolplants, Belgium. March/April. 2021. 15-20cm.

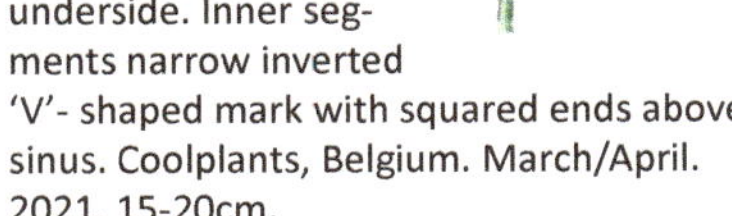

'WiFi Palatino'
Large attractive flowers with oblong green ovary. Leaves medium, erect to splayed. Outer segments broad, rounded to bluntly pointed apex, green lines above apex towards base. Inner segments broad, dark-green mark above sinus. Named for the Palatino letter style font. Coolplants, Belgium 2021. 15cm.

'WiFi Puppet on a String'
Good virescent *G. nivalis* snowdrop with long, slender flowers, heavily marked green, suspended from long, slender pedicel giving good movement. Leaves applanate, erect, slender, green. Outer segments long, slender, tapering to bluntly pointed, white apex, green mark across segment from above apex to base, narrow white margin. Inner segments dark-green mark above sinus to base, narrow white margin. Coolplants, Belgium 2022. Late February/March. 10-20cm.

'WiFi Rocket'
G. nivalis Poculiform with erect scape and spathe and long flowers suspended from slender arching pedicel beneath oblong green ovary. Leaves applanate, erect, slender, blue-green. Outer segments slender, rounded to pinched apex, lightly longitudinally ridged, green mark above apex. Inner segments green mark above sinus diffusing to base. Coolplants, Belgium. 2022.

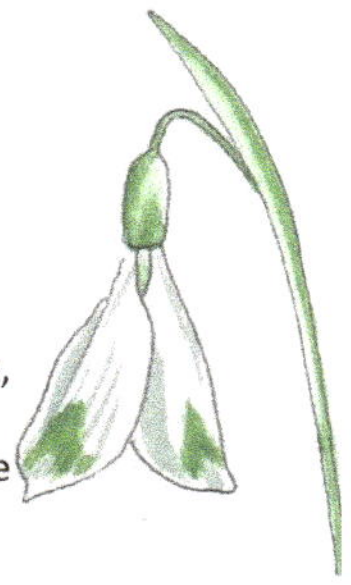

'WiFi Scha Nef'
Smaller snowdrop with good green-flushed, slender flowers, long erect pedicel and split spathe. Leaves erect to splayed, slender, green. Segments roughly of equal length, very slender, incurved, tapering to bluntly pointed apex, light-green mark above apex tapering up towards base. Coolplants, Belgium. Winter/Spring. 10cm.

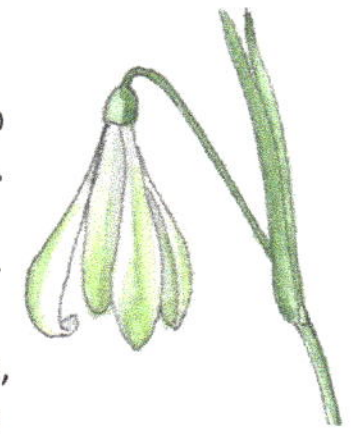

'WiFi Schacha'
Small-flowered Scharlocki type snowdrop with green-marked flowers suspended from slender ovary and pedicel beneath split spathe. Leaves arching slender, green-blue. Outer segments rounded to bluntly pointed apex, light-green wash above apex to half of segment diffusing up towards base. Inner segments, light green mark across segment above apex diffusing towards base. Discovered growing near a chateau. A 'chacha' is a well-known Belgian caramel wafer biscuit covered in chocolate. January-March. Coolplants, Belgium. 10-20cm.

'WiFi Schapeau'
Scharlockii-type hybrid with large, spreading, Poculiform flowers. Leaves slender, erect, grey-green. Segments all of equal length, slender rounded to apex, light- green mark above apex. From the French chapeau = hat. Coolplants, Belgium. 2021. 15cm.

'WiFi Shuttlecock'
Elegant *G. nivalis* with good green marks and oblong rounded ovary. Leaves applanate, slender, erect to arching, blue-green. Outer segments slender, rounded to bluntly pointed apex, lightly goffered at base, green lines above apex to half of segment, light green wash. Inner segments dark green, roughly inverted "V"-shaped mark above sinus with two arms extending upwards, paler mark towards base. Named as the shape is reminiscent of a Badminton shuttlecock. Coolplants, Belgium. March/April. 16cm.

'WiFi Sirtaki'
Elegant Inverse Poculiform flowers. Outer segments broad, flattened, rounded to apex, slightly flaring, long light-green heart-shaped mark above apex. Inner segments broad light green mark above sinus. Coolplants, Belgium. Autumn/Winter.

'WiFi Spinach Cream'
Good *G. nivalis* virescent, well-marked green. Leaves applanate, erect, slender, blue-green. Outer segments rounded, tapering to bluntly pointed apex, lightly longitudinally ridged, green lines above apex running up veins almost to base. Inner segments broad dark-green mark across segment from above sinus to base, narrow white margin. Coolplants, Belgium. 2022. Winter/Spring. 10-20cm.

'WiFi Tahoma'
Later flowering *G. nivalis* with erect, slender pedicel and strongly marked green flowers. Leaves applanate, slender, grey-green, lighter median line. Outer segments long, rounded pinched at apex, strong merging green lines above apex to half of segment. Inner segments broad, rounded, broad green mark above sinus diffusing upwards on basal side. Named for the Tahoma letter font style. Coolplants, Belgium. March/April. 15-18cm.

'WiFi Times New Roman'
Long, slender flowers suspended from erect scapes with long, slender light-green ovary. Leaves erect to arching, slender, blue-green. Outer segments, slender, tapering to pointed apex, light-green lines above apex to half of segment. Inner segments narrow, light-green inverted 'V'- shaped mark above sinus. Named for the Times New Roman letter font. Coolplants, Belgium. 2021.

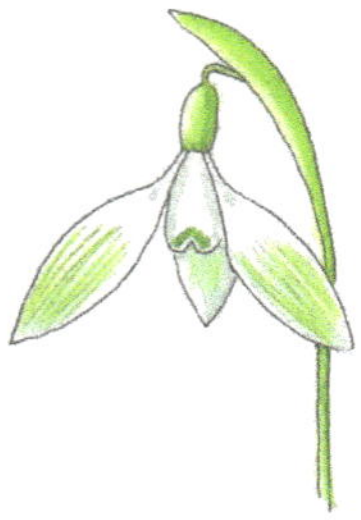

'WiFi Tipi'
Later Inverse Poculiform *G. nivalis* with small flowers and long, slender ovary. Leaves applanate, erect to arching, slender, blue-green. Outer segments clawed, rounded to rounded apex, lightly longitudinally ridged, variable green mark above apex. Inner segments narrow inverted green 'U'- shaped mark above sinus. Winter/Spring. Coolplants, Belgium. 2022. 15cm.

'WiFi Tobias'
Well-shaped green-marked flowers beneath oblong green ovary. Outer segments long, clawed, rounded to bluntly pointed apex, merging green lines above apex to half of segment. Inner segments green mark above sinus, second paler mark towards base. Coolplants, Belgium.

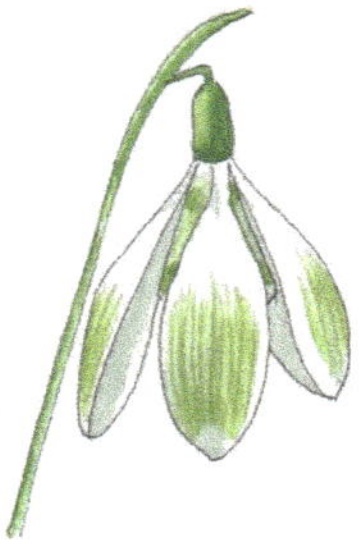

'WiFi Tombe la Neige'
Beautiful late-flowering albino *G. nivalis* snowdrop with well-shaped, large, rounded flowers. Leaves applanate, erect, slender, blue-green. Outer segments broad, short claw, rounded to bluntly pointed apex, all white. Inner segments white with large sinus, occasional green dot on either side of sinus. Named for a Belgian song sung by Adamo. Coolplants, Belgium. 2022. February/March.16cm.

'WiFi Trebuchet'
Later snowdrop with long slender flowers and oblong green ovary. Leaves grey-green, erect to arching. Outer segments boat-shaped, rounded at apex, green lines above apex. Inner segments broad dark-green mark above sinus merging into faint lines towards base. Named for the Trebuchet letter font. Coolplants, Belgium. 2021. March/April. 15-18cm.

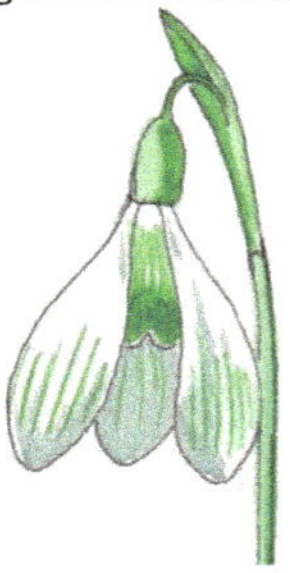

'WiFi Tripod'
Later-flowering *G. nivalis* with angular shaped flowers producing a tripod shaped appearance with long, slender, mid-green ovary. Leaves applanate, erect, slender, blue-green. Outer segments slender, rounded to pointed apex, merging green lines above apex on lower third of segment. Inner segments mid-green mark above sinus diffusing on basal side. Coolplants, Belgium 2022. March.15-18cm.

'WiFi UFO'
Attractive, Inverse Poculiform *G. plicatus* which when fully open looks similar to a flying saucer shape. Leaves explicative, medium, very green, with typical plicate folds. Outer segments broad, rounded, rounded at apex with broad green mark above small sinus. Inner segments almost as long as outer with broad green mark above sinus dividing towards base. Coolplants, Belgium 2022. February/March.

'WiFi Verdana'
Later-flowering, strongly-marked green flowers.Leaves slender, spreading, grey-green. Outer segments rounded, incurved, tapering to bluntly pointed apex. Strong merging green lines above apex. Inner segments broad 'W'-shaped mark above sinus diffusing upwards. Named for the Verdana letter font style. Coolplants, Belgium. March/April. 2021. 15-18cm.

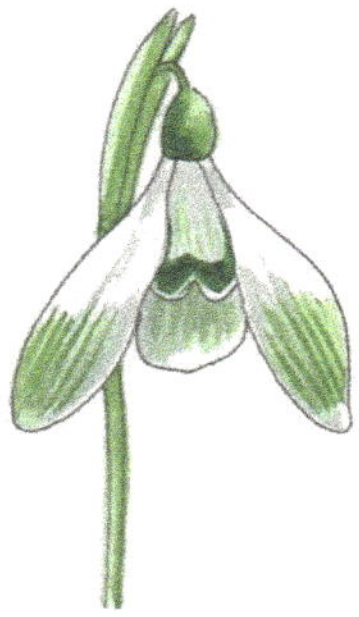

'WiFi Walk The Line'
Good green marked Virescent flowers beneath oblong, rounded green ovary.Similar in appearance to *G.* 'WiFi Borderline' but green lines on outers stretch a little further towards base and the same with the inner mark. Outer segments broad, rounded to bluntly pointed apex, spaced green lines above apex to three quarters of segment. Inner segments green mark across segment from above sinus almost to base, narrow white margin. Coolplants, Belgium. 15cm

'WiFi Wavy'
Well-marked *G. elwesii* with rounded green ovary. Leaves broad, erect to arching, grey-green. Outer segments wavy, longitudinally ridged, with green mark above rounded apex. Inner segments dark-green mark above sinus to three quarters of segment. Belgium 2022. February. 15-18cm.

'WiFi Whiplash'
Virescent *G. nivalis* with oblong, rounded green ovary. Leaves applanate, slender, erect to arching, blue-green, paler median line. Outer segments broad, rounded, incurved to rounded apex, thin olive green marks above apex to half of segment resembling the strokes of a whip. Inner segments mid-green mark above sinus diffusing towards base. Belgium 2022. February/March. 15-18cm.

'WiFi White Spot'
Well-marked virescent *G. nivalis* with slender, oblong mid-green ovary. Outer segments slender, rounded tapering to bluntly pointed apes, mid-green lines above apex towards base, paler green wash. inner segments green mark above sinus diffusing slightly towards base, with paler central spot, hence name. Belgium 2022. February/March. 15-18cm.

'WiFi With Flair'
Very elegant, late flowering *G. nivalis* snowdrop with large bell-shaped flowers and oblong, rounded green ovary. Leaves applanate, medium, erect to arching, grey-green, paler median line. Outer segments slender, rounded, incurved to pinched apex, green lines above apex on lower third with paler wash. Inner segments inverted green 'V'- to 'W'-shaped mark above sinus. Belgium 2022. March. 15-20cm.

'WiFi Yellow Cherry'
Good combination of green and gold. Green spathe with yellow pedicel and slender yellow ovary. Outer segments slender, incurved, pointed at apex. Inner segments good yellow inverted 'V'- shaped mark with rounded ends above sinus. Coolplants, Belgium.

'Windy Ridge'
An apparent hybrid between *G. plicatus* and *G. elwesii* with very blue-green spathe and foliage, thought to have originated at Windyridge Nursery, Dublin. Leaves broad, erect to arching, very blue-grey in colour. Rounded green ovary. Outer segments short claw and pronounced shoulder, rounded, incurved to pointed apex. Inner segments green mark above sinus merging into second mark towards base.

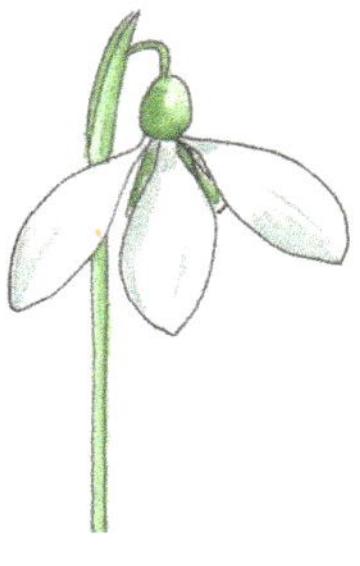

'Winter Green'
A rare *G. plicatus* with very green foliage and lightly textured flowers. Leaves explicative, broad, shiny, very green with grooved margins. Inner segments green mark above sinus. Raised by John Morley, North Green Snowdrops and grown in their garden for many years.

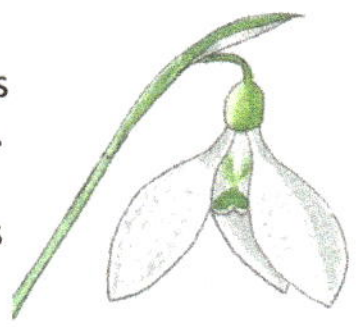

'Winter Memory'
G. nivalis Pocuiform with rounded flowers. Leaves applanate, erect, slender, blue-green. Segments all of equal length, slender, rounded to bluntly pointed apex, marked green above apex. Oliver Vico. 2019. Winter/Spring.

'Withnail'
G. nivalis virescent snowdrop with long, slender, green marked flowers. Leaves applanate, erect, slender, blue-green. Outer segments long, slender, rounded and tapering towards bluntly pointed apex, small green lines above apex. Inner segments broad inverted 'U'- to 'V'-shaped mark above sinus. Decora Nursery, Croatia. Winter/Spring.

'Witton Dumpy'
Shorter-growing snowdrop with large, well-shaped, rounded flowers. Leaves erect to arching, slender, blue-green. Outer segments broad, slender claw, incurved, rounded to bluntly pointed apex. Inner segments narrow green inverted 'V'- shaped mark above sinus. Richard Hobbs.10cm.

'Wol's Tubby' (Syn GC59)
Very large, rounded, heavily textured and puckered flowers. Outer segments broad, rounded, puckered. Inner segments green mark above sinus to base. Named for Will Staines as a reference to his constant battle with his waistline. Wol and Sue Staines, Glen Chantry. 16cm.

'Wonder Green'
G. nivalis with long, slender, green marked flowers and oblong green ovary. Leaves applanate, erect, slender, blue-green. Outer segments long, very slender, incurved to pointed apex, lightly longitudinally ridged, light-green shading across segment from above apex. Inner segments mid-green mark above sinus almost to base, narrow white margin. Domaine du Ooievaar Nursery, Flanders, Belgium. 2021. Winter/Spring.

'X-Ray'
Elegant, long, slender-looking flowers. Leaves medium, erect to arching, grey-green, paler median line. Outer segments long slender claw, lightly longitudinally ridged, tapering to pointed apex. Inner segments short, flattened green inverted 'U'- shaped mark above sinus. 2015.

'Yellow Dangly'
An Italian seedling which is either a Hybrid or a yellow *G. reginae-olgae*. Elegant flowers suspended from long pedicel. Ian Christie. 2018.

'Yuletide'
A *G. elwesii* cultivar with slender looking, early flowers. And oblong green ovary Leaves supervolute, broad, pointed, arching, grey-green, well developed at flowering. Outer segments slender, incurved to pointed apex. Inner segments good heart-shaped green mark above sinus to half of segment. Phil Cornish from E. B. Anderson's former garden, Lower Slaughter, Gloucestershire. 1985. Flowers around Christmas. 17cm.

'Yvette Camble'
An American-selected *G. nivalis*. Leaves applanate, slender, blue-green, splayed. Outer segments boat-shaped, rounded to bluntly pointed apex, merging green lines above apex to half of segment. Inner segments broad green mark above sinus. Edgwood Gardens, USA. 10-15cm.

'Zagreb'
Small, neat *G. nivalis*. Leaves applanate, erect to arching, slender, blue-green-glaucous. Outer segments slender, rounded to pointed apex. Inner segments narrow green inverted 'V'-shaped mark above sinus. Bulks up well. Originating in Zagreb.

'Zakim'
G. gracilis with well rounded balloon like flowers. Leaves, applanate, broad and characteristically twisted. Outer segments very broad, rounded to bluntly pointed apex. Inner segments crisp dark green, bridge-like mark above sinus and second broad mark towards base. Named for the Leonard P. Zakim Bunker Hill Memorial Bridge in Boston. Rick Goodenough, USA from seed collected by Tom Mitchell near Izmir, Turkey and grown on by Lonsdale Gardens, USA. Grows well.

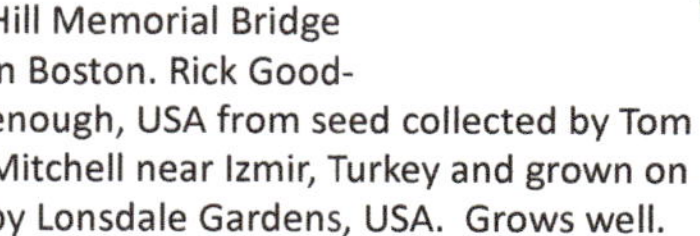

'Zeitgeist'
G. plicatus with handsome long flowers suspended from erect, slender pedicel. Leaves explicative, erect, plicate, green. Outer segments long claw, rounded to bluntly pointed apex, longitudinally ridged, four to five short green lines above apex with light wash. Inner segments long to half of outer, broad mid-green horseshoe shaped mark above sinus. Winter/Spring. 16cm.

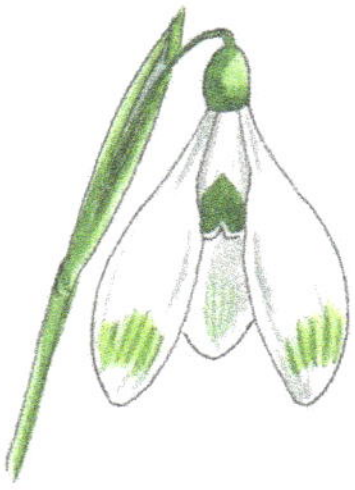

'Zunzuncito'
G. nivalis Inverse Poculiform snowdrop with very white flowers, small green marks and oblong green ovary. Leaves applanate, erect to arching, blue-green. Outer segments, broad, flattened, rounded at apex, small green mark above small sinus. Inner segments small green mark above sinus.

LATE ADDITIONS

'Eden'
Similar looking snowdrop to *G.* 'Robin Hood', with flowers suspended from slender pedicel on tall, erect scapes. Leaves slender, erect, blue-green-glaucous. Outer segments long, clawed, rounded, tapering to pointed apex. Inner segments green mark across segment from above sinus. Good, strong grower. Probably hybrid between *G. plicatus* and *G. elwesii*. Found by Bill Clarke, Wandlebury, Cambridgeshire and named for his grandson. Slow to increase. Winter. 20cm.

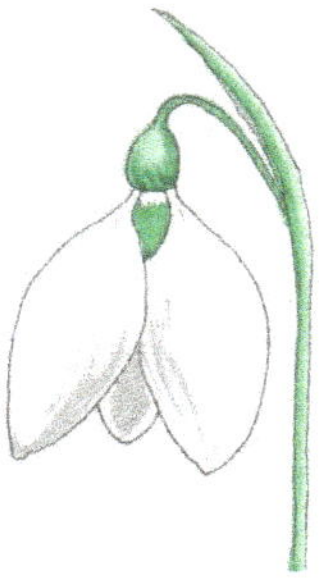

'WiFi Cambria'
G. nivalis well-shaped flowers and slender green ovary. Leaves applanate, slender erect to arching. blue-green. Outer segments clawed, rounded to pointed apex, green lines above apex to half of segment. Inner segments inverted green 'U'- shaped mark above sinus merging into green marks extending to base leaving a white centre. Coolplants, Belgium. 2021. 15cm.

Alphabetical List of Recently Named Snowdrops in this Directory

Recently Named Snowdrops in this Volume

**Indicates that the variety is not represented by an image in the Directory*

1. Adam
2. Adamite
3. Ade-Le
4. Adriana
5. Aero*
6. Alaina
7. Alan's Long Ovary
8. Alban Arthan
9. Albatross
10. Alexander Beryl
11. Alexandrite
12. Alice's Late
13. Alles im Grunnen Bereich
14. Alpha Apricot Ice Cream
15. Alpha Autumn Miracle
16. Alpha Bermudian Triangle
17. Alph Boiling Milk
18. Alpha Bonus
19. Alpha Boo
20. Alpha Cat's Toy
21. Alpha Children of Triffids
22. Alpha Chinese Lantern
23. Alpha Christmas Bells
24. Alpha Compliment
25. Alpha Dachshund
26. Alpha Detective Poirot
27. Alpha Droopy
28. Alpha Elhaz
29. Alpha English Lawn
30. Alpha First Snow
31. Alpha Flying Birds
32. Alpha Gold n Pearls
33. Alpha Guppies
34. Alpha Helmet (Tutonic Helmet)
35. Alpha Hope
36. Alpha Humpty Dumpty
37. Alpha Husky
38. Alpha Icicles
39. Alpha Jade Comb
40. Alpha Kiwi Mousse
41. Alpha Ladybug
42. Alpha Light Nostalgia
43. Alpha Little Mermaid
44. Alpha Merlin's Brushes
45. Alpha New Art
46. Alpha Old Joke
47. Alpha Paradox
48. Alpha Pistachio Ice Cream
49. Alpha Point
50. Alpha Praying Hands
51. Alpha Predator
52. Alpha Pug Dog
53. Alpha Roly Poly
54. Alpha Seven Dwarfs
55. Alpha Spring Crinoline
56. Alpha Striped Beauty
57. Alpha Striped Skittles
58. Alpha Suns Favourite
59. Alpha Tutelar
60. Alpha White Claws
61. Alpha White Dragon
62. Alpha Yellow Fairy
63. Alpha Yeti
64. *alpinus* var. *alpinus*
65. Altered Image
66. Amy Doncaster's Double
67. Angel Fley
68. Angel Lady
69. Angel of the North
70. Angel or Demon
71. Angela's Gift
72. Anglesey Cloudgazer
73. Anglesey Doublet
74. Anglesey Promise
75. Ann Baring
76. Ann Christin
77. Annalivia
78. Anna O
79. Anne's Pyjamas
80. Antares
81. Apatite
82. Apollo
83. Appleby Spiky
84. Arabian Gold
85. Ariadne
86. Arto's Secret
87. Arachnophobia
88. Ate Tea
89. Atkinsii Variegated
90. Augmented Triad
91. August Der Starke
92. Aul
93. Aunt Nellie Danglers
94. Aztec Gold
95. Babraham Braveheart
96. Bad Hair Day
97. Badgers End
98. Ball

99. Bambino
100. Barleymow
101. Basket*
102. Be My Valentine
103. Bead Bear
104. Beekeeping Cream of Bauerlein
105. Belgium Trust
106. Belisha Beacon
107. Bell Star
108. Bella Bianco
109. Bend Bear
110. Beta Galanthopath
111. Betelgeuze
112. Bethany
113. Betty Hughes
114. Big Ben
115. Big Egg
116. Big One
117. Bitter Lemons
118. Bledlow Double*
119. Blond Birgit
120. Blonde Isabel
121. Blue Jay
122. Blue Octopus
123. Boat Bear
124. Bob's Seedling
125. Bonanza
126. Bong Bear
127. Bowles Boys
128. Bowles Gemini
129. Bowling Bear
130. Brechin Tower
131. Brian Ellis
132. Broad Leaved Convolute
133. Bullfight
134. Bunzaue
135. Burley Bear
136. *bursanus*
137. *bursanus* Green Tip
138. Buttons and Bows
139. Cal-U-Met
140. Calabrian Green
141. Cape Cod
142. Captain Dunstan
143. Captain Hook
144. Carinthian Green Giant
145. Carlos aus Santana
146. Carol Dancer
147. Castle Green
148. Castle Green Shadow
149. Castle Little Star*
150. Castle New
151. Castle The Maules
152. Caterpillar
153. Catherine McCauley
154. Celia Meggers
155. Celia's Double
156. Chadwick's Cream
157. Chantry Bonus
158. Chantry Constable
159. Chantry Cracker
160. Chantry Dame
161. Chantry Dougal
162. Chantry Duchess
163. Chantry Empress
164. Chantry Gold Cross
165. Chantry Lady
166. Chantry Ladylike
167. Chantry Pinstripe
168. Chantry Poppet
169. Charley Barr*
170. Charlie's Angel
171. Charmer-Flore-Pleno
172. Che Bel Fior
173. Chicken Pen*
174. Child of Forest
175. Chiltern Queens
176. Chimaera
177. Chorus Line
178. Chthonic
179. Chubby
180. Clemens Heidger
181. Clun
182. Clun Green Convolute
183. Cockapoo
184. Cocoon
185. Comet's Tail
186. Compton Green Eye
187. Constellation
188. Corovode*
189. Cowichan Caribou
190. Cream Bear
191. Cressida
192. Crime Wild
193. Cross of Lattera
194. Crown Jewel
195. Cup Of Joy
196. Curse of the Were Rabbit
197. Curtsey
198. Daft Punk
199. Dalkleid
200. Dangling Green
201. Danish Gold
202. David Foreman
203. Divine Comedy*
204. Dizzy
205. Doesn't Must
206. Domaine du Oojevaar
207. Double Decker
208. Double Dragon
209. Double Hedgehog*
210. Double Orange Star
211. Double Vision
212. Dr Rodgerson's No 2 Hybrid
213. Dragon Gold
214. Dryad Blizzard
215. Dryad Cerberus
216. Dryad Demeter
217. Dryad Echo
218. Dryad Gold Sceptre
219. Dryad Gold Standard
220. Dryad Hera
221. Dryad Jupiter
222. Dryad Terpsichore
223. Dryad Venus
224. Dryad Zeus
225. DSC-3066 Tuuliku

226. Duizendduin
227. Dumbo
228. Early Autumn Jewel
229. Early Green
230. Easter Parade
231. Echoes*
232. Eden
233. Edward Ray Robie
234. Eeyore
235. Eifeler Grün
236. El Bandito
237. El Mono
238. Elegant Bear
239. Elizabeth Benfield
240. Ellie Saunders
241. Elven Lantern
242. Emerald Bear
243. Emerald Pagoda
244. Eppie
245. Equinox
246. Eric Covell
247. Eric Saints
248. Eric Watson
249. Ernie Cavallo
250. Essex Girl
251. Etoile des Neiges
252. ex Cappoquin-House
253. ex Cherub seedling
254. ex Gedelme
255. ex Gert Geenson
256. ex Montrose
257. ex Ted Kiely
258. Eye Eye
259. Fairlight
260. Fanfreluche
261. Fast Zerdatscht
262. Feline
263. Fieldgate Fantasia
264. Fieldgate Strong
265. Fifi
266. Fireside Find
267. Flame
268. Flanders Beauty
269. Flanders Field
270. Flavia
271. Floriferous Star
272. Forest Muse
273. Forger's Brush
274. Four Play
275. Frail Beauty
276. Freak
277. Fred's Surprise
278. Freddy King
279. Freddy's White Wonder
280. Fridge Magnet
281. Frohnauer Gold
282. Full Marks
283. Fuzzy Wuzzy*
284. Gaia
285. Galloping Horses
286. Garden Peace
287. Gekke Henkie
288. Genainville
289. Georgia Arula
290. Georgiana
291. Gestrichelt Pulk
292. Gimley Din
293. Gimley Durin
294. Glasdrum
295. Glauconite
296. Glen Una
297. Glooming
298. Glory Gold
299. Glowing
300. Gold 'n' Green
301. Golden Acre
302. Golden Eagle
303. Golden Girl
304. Golden Goose
305. Golden Plummet
306. Golden Tears
307. Golden Wings
308. Goldfinch
309. Goldfinger
310. Goldheart
311. Goldilocks
312. Goldish
313. Goldsmith
314. Gore Booth
315. Gorkains
316. Gotborg
317. Gothenburg Double
318. *gracilis* Kew Form
319. Grake's Oddity
320. Green Bear
321. Green Besom
322. Green Bonnet
323. Green Dance
324. Green Day
325. Green Desire
326. Green Early Jewel
327. Green Fairy
328. Green Flight
329. Green Genes
330. Green Hell
331. Green Jewel*
332. Green Lampshade
333. Green Longfinger
334. Green Moon*
335. Green Peacock
336. Green Spring Fling
337. Green Testament
338. Green Tiger
339. Green Tipped Balloon
340. Green Tipped Orange Star
341. Green Tornadoes
342. Green Trym
343. Green Whisp
344. Greengage
345. Greenpeace
346. Grizzly
347. Grossenhof
348. Grumpy's Brother
349. Grüne Donauzwerg
350. Grüne Heike
351. Grüne Nase
352. Grüne Spitze
353. Grünspecht No 4
354. GRY2
355. Gwawis Caribou
356. Haconby Early
357. Halcyon

358. Hanging Out
359. Hardwick
360. Harley Milne
361. Haroula
362. Harpur Grace
363. Hattrick
364. Hawk
365. Heaven's Patch
366. Heloise des Essourts
367. Henley Deep Green
368. Herzilein
369. Hidden Secret
370. High Tide
371. Hill View
372. Honeypie
373. Howick Starlight
374. Ice Princess
375. Italian Yellow
376. Jack By The Edge
377. Janet Cropley
378. Jasper
379. Je Suis Charlie
380. Joe's Poculiform
381. John Morley
382. John Owen
383. Judith
384. Julius Scholz
385. Just White
386. Kaina
387. Kanari
388. Katherine James
389. Kato
390. Karst Dumpy*
391. Kaur
392. Kelpie
393. Kencot Pickles
394. Kera
395. Kermode Bear
396. Ketskhoveli
397. Kill Abbey
398. Kiur
399. Koksilah Caribou
400. Kokoshnik Diadem
401. Komma
402. Krim Wild
403. Kudrus
404. La Mornier
405. La Vie En Rose
406. Lady MacBeth
407. Ladyboy
408. Lagle
409. Lamplighter
410. Lanky
411. Leilani
412. Lemon
413. Lemon and Lime
414. Lemon Kindness
415. Liebsten
416. Lilliput
417. Lil'wat Caribou
418. Lili Vegrin
419. Lilliput
420. Limetime
421. Limetree
422. Lin
423. Lincoln Green
424. Little Angel
425. Little Artist
426. Little Beauty
427. Little Bell
428. Little Green Grasshopper
429. Little Prince
430. Lola
431. Lonely Island
432. Longbow
433. Longheart
434. Longwood Green Tip
435. Look At Me
436. Look At Yourself
437. Louella
438. Louise
439. Magic
440. Magnetic Pedicel
441. Maja's Gift
442. Margaret Ford
443. Margaret Markham
444. Margaret Mitchell
445. Marion's Choice
446. Marion's Surprise
447. Marman's Emerald
448. Marmot Chartreuse
449. Marmot Cross
450. Marmot Duncan
451. Marmot Flight
452. Marmot Ghost
453. Marmot Koksilah
454. Marmot Puckered
455. Marmot Sage Grouse
456. Marmot Shocker
457. Marmot Splits
458. Marmot Stripe
459. Marmot Teardrop
460. Marmot Two
461. Maryse's Double
462. Marzenbecher
463. Maulsden
464. Medena
465. Melody in Green
466. Meltsas
467. Merlin Bonds Form
468. Merlins Herbstgefoige
469. Meteor
470. Mett
471. Mex
472. Midori
473. Mila
474. Milch und Honig
475. Minerali Bani*
476. Mini Poculiformis
477. Minions Smiling
478. Miss Tilly
479. Misty Jewel
480. Mita
481. Mmm
482. Moby Dick
483. Money
484. Montrose
485. Moonlight Shadow
486. Moonshine
487. Moose Bear
488. More Than A Feeling
489. Moses Autumn Jewel
490. Motabile

491. Mr Camoflage
492. Mr. Courage's Early
493. Mr Frayling's Double
494. Mr Peggoty
495. Mr. Positive
496. Mr. T. Mackie
497. Mr Taylor
498. Mutant
499. Neat Green
500. Nellie Brinsley's Double
501. Nelle-e-Belle
502. Nelly's Birthday
503. Neon Kiss
504. Nessie
505. New Art
506. Norma
507. North Green Wasp
508. Nutt's Early
509. Odd Sharlock
510. Of*
511. Okidoki
512. Olana
513. Old Kite
514. Olivia Newton John
515. Orange Star
516. Orion
517. Ostergras
518. Otterhauw*
519. Paleface
520. Papageno
521. Parakeet
522. Paranoid
523. Parfetta
524. Party Dress
525. Pear Bear
526. Peg's Double
527. Peppermint Candy
528. Pepys
529. Peter Davis
530. Petra
531. Petrich*
532. Phoebe
533. Pik Dame
534. Pil Pil
535. Pingpoc
536. Piping Plover
537. Praying Hands
538. Priscilla's Green
539. Puffin
540. Pumpkin
541. Punky
542. Quasimodo
543. Rachel Mahaffy
544. Rainbow Alcyone
545. Rainbow Balthazar
546. Rainbow Bealvi
547. Rainbow Caspar
548. Rainbow Christmas Fairy
549. Rainbow Golden Girl
550. Rainbow Gold Top
551. Rainbow Green Gracie
552. Rainbow Melchior
553. Rainbow Selene
554. Rainbow's End
555. Rapunzel
556. S.F.On The Move
557. Reflection
558. Renata Meier
559. Richard Nutt's Green Leaved Hybrid
560. Richard Todd
561. Ristpart
562. Robust Timpany
563. Rocket Ship
564. Rockland
565. Rodmarton
566. Roll On Jo
567. Rose Baron
568. Rosie
569. Roy Buchanan
570. Roy's Double
571. Ruby's Green Find
572. Rus Ukraine
573. Ruslan
574. S.F. On The Move
575. Salad Bowl
576. *samothracicus*
577. Sandalio's Charm
578. Sanger
579. SA0901
580. Sarah's Sweet Heart
581. Scaramouche
582. Scharlockshank
583. Schewefelgold
584. Schwefeifee
585. Schorbuser Maze
586. Seratin
587. Seremony
588. Shampita
589. Shimmer
590. Shrubbery Special
591. Sibbertoft Soldier
592. Silver Star
593. Simone
594. Sioux
595. Sixtus
596. Snow Angel
597. Snow Flurry
598. Snowhite ex Hulsman
599. Solo
600. Spare Bear
601. Spili
602. Spokes Pinwheel
603. Spring Cottage
604. Sprite
605. Squidley
606. St Antonius
607. Starbright
608. Stiebritz (see Bull fight)*
609. Stork
610. Strasbourg Twins
611. Streets Ahead
612. Streifenzart
613. Strewelpeter
614. Stuard Turner
615. Styx
616. Sunbeam
617. Sunlime
618. Sunrise
619. Svelt
620. Sweet Gaia

621. Swinging Star
622. Sylvan
623. Tangfastic
624. Tante Anne
625. Tatiana Superb
626. TCH-0110
627. The Artist
628. The Green Bucket
629. The Lady With the Lamp
630. The Paddle
631. The Stripper
632. Timeless Teresa
633. Timmy Whiteley
634. Timpany Bold
635. Tinus Mustache
636. Tinus Snor
637. Tiny Sunshine Glow
638. Tom's Fool
639. Torgian Geut
640. Tramlines
641. Trinity
642. Trudy Spotted
643. Truffle
644. Trumpses
645. Trymskrams
646. Tsoe-Tsoe Mama
647. Tuff
648. Tumbledown
649. Tuppence
650. Tutkas
651. Tuuliku 8
652. Twilight
653. Two Tone
654. Uli's Gift
655. Uranium
656. Uwe Wagner
657. Vart
658. Veliki Progasti
659. Vlaam Re
660. Walkden Spiky*
661. Walnut
662. Walter's Double
663. Waltham Place 2
664. Wantage Green
665. Warwickshire Gemini
666. Wassergeist
667. Watersmeet
668. Watlington Green-man
669. Watlington Smudge
670. Watson L
671. Well Hung
672. Wessex Advent
673. Wessex Head
674. WeWei Caribou
675. Whirling Dervish
676. White Gold
677. Whiteheart
678. White Walker
679. Widder
680. WiFi Aquarel
681. WiFi Arial
682. WiFi Bahnschrift
683. WiFi Benji
684. WiFi Big Brother
685. WiFi Bingo
686. WiFi Borderline
687. WiFi Calibri
688. WiFi Cambria
689. WiFi Caret
690. WiFi Century
691. WiFi Comic Sans
692. WiFi Dipping
693. WiFi Dotty
694. WiFi Eden's Basket
695. WiFi Flair
696. WiFi Flight of Hearts
697. WiFi Green Bride
698. WiFi Kiruha
699. WiFi Kittiwake
700. WiFi Let's Dance
701. WiFi Lucida
702. WiFi Magic Mint
703. WiFi Monroe
704. WiFi Myriad
705. WiFi Nacitti
706. WiFi Nigeria
707. WiFi Overflow
708. WiFi Palatino
709. WiFi Puppet on a String
710. WiFi Rocket
711. WiFi Scha Nef
712. WiFi Schach
713. WiFi Schapeau
714. WiFi Shuttlecock
715. WiFi Sirtaki
716. WiFi Spinach Cream
717. WiFi Tahoma
718. WiFi Times New Roman
719. WiFi Tipi
720. WiFi Tobias
721. WiFi Tombe la Neige
722. WiFi Trebuchet
723. WiFi Tripod
724. WiFi Ufo
725. WiFi Verdana
726 WiFi Walk the Line
727. WiFi Wavy
728. WiFi Whiplash
729. WiFi White Spot
730. WiFi With Flair
731. WiFi Yellow Cherry
732. Windy Ridge
733. Winter Green
734. Winter Memory
735. Withnail
736. Witton Dumpy
737. Wol's Tubby
738. Wonder Green
739. X-Ray
740. Yellow Dangly*
741. Yuletide
742. Yvette Camble
743. Zagreb
744. Zakim
745. Zeitgeist
746. Zunzuncito

Galanthus
Names Currently in Circulation

Galanthus Names Currently in Circulation

1. Abington Green
2. Acton Pigot No 1
3. Acton Pigot 2
4. Acton Pigot No 3
5. Adam
6. Ade-Le
7. Adriana
8. Advent
9. Advolly Richmond
10. Aestivalis
11. Aestivum
12. Afterglow
13. Aidin
14. Aioli
15. Ailwyn
16. AJM 75
17. Alaaf
18. Aladdin
19. Alaina
20. Alan Clark
21. Alan's Long Ovary
22. Alan's Treat
23. Alanya Yayla
24. Alban Arthan
25. Albatross
26. Alburgh Claw
27. Albus
28. Aldgate 75
29. Aldo Nefat
30. Alex Duguid
31. Alexander the Great
32. Alexandra
33. Alexandrite
34. Aliana
35. Alice Bell
36. Alice's Late
37. Alison Hilary
38. Allen's Perfection
39. Allen's seedling
40. All Saints
41. Almost All Green
42. Alpha Apricot Ice Cream
43. Alpha Autumn Miracle
44. Alpha Bird's Beak
45. Alpha Blush of Spring
46. Alpha Boo
47. Alpha Children of Triffids
48. Alpha Children's Laughter
49. Alpha Chinese Lantern
50. Alpha Christmas Angel
51. Alpha Chupakabra
52. Alpha Cocker Spaniel
53. Alpha Dark Heart
54. Alpha Dervish Dance
55. Alpha Droopy
56. Alpha Elegant Turn
57. Alpha Emerald Bells
58. Alpha Fountain
59. Alpha Green Bees
60. Alpha Green Chaos
61. Alpha Green Cheeks
62. Alpha Green Dawn
63. Alpha Green Fog
64. Alpha Green Lantern
65. Alpha Green Miracle
66. Alpha Green Peacock
67. Alpha Green Veil
68. Alpha Happy Duckling
69. Alpha Hope
70. Alpha Jade Comb
71. Alpha Kingfisher
72. Alpha Kiss of Spring
73. Alpha Light Nostalgia
74. Alpha Little Mermaid
75. Alpha Magic Chryso prase
76. Alpha Masterpiece of Spring
77. Alpha Meridians
78. Alpha Microball
79. Alpha Olga
80. Alpha Pierrot
81. Alpha Pug Dog
82. Alpha Sassoon
83. Alpha Sergey Nigoyan
84. Alpha Shell
85. Alpha Snow Bugs
86. Alpha Snow Dragon
87. Augmented Tryad
88. Alpha Spring Balloon
89. Alpha Spring Symphony
90. Alpha Striped Beauty
91. Alpha Tatjana
92. Alpha Tenderness
93. Alpha White Dragon
94. Alpha White Wavy Vase
95. Alpha Yellow Fairy
96. Alpha Yuri Verbitsky
97. *alpinus*
98. *alpinus* var. *alpinus*
99. *alpinus* bortkewen-sianus
100. Altered Image

101. Altheia
102. Amberglow
103. Amelia
104. Amigo
105. Amy Doncaster
106. Amy Doncaster's Double
107. Amy Jade
108. Anacreon
109. Anamus
110. Andreas Fault
111. Andrew Thorpe
112. Angel or Demon
113. Angel of the North
114. Angel Lady
115. Angela's Early
116. Angelfly
117. Angelina
118. Angelique
119. Angie
120. Anglepoise
121. Anglesey Abbey
122. Anglesey Adder
123. Anglesey Aurora
124. Anglesey Candlelight
125. Anglesey Doublet
126. Anglesey Not Galatea
127. Anglesey Orange Tip
128. Anglesey Promise
129. Anglesey Rainbow
130. Anglesey Spiky
131. Angustifolius
132. Anika
133. Anita
134. Anmarie Kee
135. Anmut
136. Ann-Christin
137. Ann of Geierstein
138. Ann's Millenium Giant
139. Anna
140. Anne
141. Anneke Classen
142. Annette
143. Annielle
144. Antares
145. Antrym
146. Antwerp Greentip
147. Aphrodite
148. Apfelgrun
149. Applause
150. Appleby
151. Appleby One
152. Appleby Spiky
153. April Fool
154. April Tutu
155. Arabian Gold
156. Arachnofobia
157. Aragnee
158. Aragorn
159. Ardens Silence
160. Ariadne
161. Armilde
162. Armine
163. Armistice Day
164. Arnold
165. Artemis
166. Art Nouveau
167. Artjuschenkoae
168. Ate Tea
169. Athenae
170. Atkinsii
171. Atkinsii of Finnis
172. Atkinsii Moccas Form
173. Atlas
174. Audry Vockins
175. Augenpaar
176. Augenschmaus
177. Augenweide
178. Augmented Triad
179. August
180. Augustus
181. Augmented Tryad
182. Aunt Agnes
183. Aunt Nellie Danglers
184. Aunty Flo
185. Aura
186. Aurelia
187. Aurora
188. Auslese Elbe
189. Autumn Beauty
190. Autumn Belle
191. Autumn Snow
192. Ayres Rock
193. Ayes and Noes
194. Babraham
195. Babraham Dwarf
196. Babraham Scented
197. Baby Arnott
198. Baby Belle
199. Backhouse Spectacles
200. Bagpuize Alexander
201. Bagpuize Elizabeth
202. Bagpuize Frances
203. Bagpuize Greentip
204. Bagpuize Virginia
205. Bagpuize Walrus
206. Ballard
207. Ballard's Green Hybrid
208. Ballard's No Notch
209. Ballerina
210. Balloon
211. Ballynahinch
212. Bankside
213. Barbara Buchanan's Late
214. Barbara Dibley
215. Barbara's Double
216. Barbara's Hybrid
217. Barguest
218. Barnes
219. Barnhill
220. Baroness Ranson-ette
221. Baselflecken
222. Basisgrüner
223. Batmobile
224. Baxendale's Late
225. Baylham
226. Baytop
227. Bead Bear
228. Beany
229. Beatrice
230. Beechwood
231. Beenak
232. Beethoven
233. Beetle

234. Belated
235. Belgium Trust
236. Beli Zajcek
237. Bella
238. Belle Bianco
239. Belle de Wallonie
240. Belle Etoiles
241. Belle Etoile de Maritime
242. Bell Star
243. Belle Therese
244. Beloglavi
245. Beluga
246. Belvedere Gold
247. Ben Warwick
248. Be My Valentine
249. Benedict
250. Benhall Beauty
251. Benhall Seedling
252. Benjamin Britten
253. Bentley Green
254. Benton Magnet
255. Berkeley
256. Berliner Kaiser
257. Bernard Röllich
258. Bernard Tickner
259. Bertha
260. Berthilla
261. Bertram Anderson
262. Berwick
263. Bess
264. Betelgeuze
265. Beth Chatto
266. Betty Frazer
267. Betty Hamilton
268. Betty Hansell
269. Betty Hughes
270. Betty Page
271. Biflorus
272. Big Ben
273. Big Bertha
274. Big Bopper
275. Big Boy
276. Big Egg
277. Big Eyes
278. Big Mask
279. Big One
280. Big Star
281. Bill Baker's Early
282. Bill Baker's Green Tipped
283. Bill Baker's Large
284. Bill Baker's Maximus
285. Bill Bishop
286. Bill Boardman
287. Bill Clark
288. Billinghurst
289. Bingljajoci Bumerang
290. Biscapus
291. Bishop's Mitre
292. Bitter Lemon
293. Bitton
294. Black Magic
295. Blackthorn
296. Bladerunner
297. Blanc de Chine
298. Blaris
299. Blasses Wesen
300. Blattgrün
301. Blau-Weis Schorbus
302. Bletsoe Castle
303. Blewbury
304. Blewbury Tart
305. Blithe Spirit
306. Blonde Andrea
307. Blond Betty
308. Blond Birgit
309. Blonde Erika
310. Blonde Inge
311. Blonde Isabel
312. Blonde Ommerschans
313. Bloomer
314. Blue Jay
315. Blue John
316. Blue Octopus
317. Blue Peter
318. Blue Streak
319. Blue Trym
320. Blur
321. Boat Bear
322. Bobette
323. Bob Nelson
324. Body's Green
325. Bob's Seedling
326. Boge
327. Bogenlampe – Ark Lamp
328. Bohemia Gold
329. Bohemia Skirt
330. Bohemia White
331. Bohemian Rhapsody
332. Bolu Shades
333. Bonanaza
334. Bong Bear
335. Boromir
336. Boschhoeve
337. Bowles Boys
338. Bowles Gemini
339. Bowles Large
340. Bowle's Late
341. Boydii
342. Bowling Bear
343. Boyd's Double
344. Brandling
345. Brechin Ian
346. Brechin Tower
347. Brenda Troyle
348. Brest
349. Breviflos
350. Brian Ellis
351. Brian Mathew
352. Brian Spence
353. Bridesmaid
354. Bridgham
355. Brigadier Mathias
356. Bright Eyes
357. Brinn
358. Britta
359. Britten's Kite
360. Broadwell
361. Broad Leaved Convolute
362. Brocklamont Seedling

363. Broske
364. Bruce Lloyd
365. Brumptons Freak
366. Bryan Hewitt
367. Bubble
368. Bucks Green Tip
369. Budenzauber
370. Bullfinch
371. Bumble Green
372. Bumblebee
373. Bunch
374. Bungee
375. Burford House
376. Burly Bear
377. *bursanus*
378. Bushmills
379. Button
380. Butt's Form
381. Buzet
382. By Gate
383. Byfield Special
384. Calabria
385. Calabrian Green
386. Calebasse
387. Calvors
388. Cambridge
389. Camouflage
390. Canadian Winter
391. Candidus
392. Cape Cod
393. Caprea 115
394. Caprea 119
395. Carina
396. Carinthian Elronge
397. Carinthian Green
398. Carinthian Green Giant
399. Carinthian I-Poc
400. Carinthian Spring
401. Carinthian Summer
402. Carinthian Sun
403. Carinthian War Bonnett
404. Carlos Santana
405. Carol Dancer
406. Carol Simcoe
407. Caroline
408. Caroline Elwes
409. Carpathian Giant
410. Carpathian Viridapicis
411. Carpentry Shop
412. Caryl Baron
413. Casa Nova
414. Casanovas Heim kehr
415. Caspar
416. Cassaba
417. Cassaba Boydii
418. Castle Green Shadow
419. Castle Early
420. Castle Eye Shadow
421. Castle Eyes
422. Castle Giant
423. Castle Green
424. Castle Green Dragon
425. Castle Lemon
426. Castle New
427. Castle Peardrop
428. Castle Plum
429. Castle Select
430. Castle Tall
431. Castle the Maules
432. Castle Twin Head
433. Castlegar
434. Cathcartiae
435. Caterpillar
436. Catherine Boardman
437. Catherine McCauley
438. Caucasicus
439. Causpicus
440. Cedric's Prolific
441. Celadon
442. Celia Blakeway-Phillips
443. Celia Meggers
444. Celia Sawyer
445. Celia's Double
446. Cerberus
447. Ceri Roberts
448. Cerkno
449. Cesta
450. Chadwicks Cream
451. Chameleon
452. Chandlers Green-Tipped
453. Chantry Cracker
454. Chantry Gold Cross
455. Chantry Green Twins
456. Chantry Lady
457. Chantry Poppet
458. Chantry Taffeta
459. Chapeli
460. Charles Wingfield
461. Charlie
462. Charlotte
463. Charlotte Jean
464. Charmer
465. Charmer Flore Pleno
466. Chatterbox
467. Chatton
468. Chedworth
469. Chel Bel Fior
470. Chelsworth Magnet
471. Chequers
472. Cherry Bomb
473. Cherub
474. Chetwode Greentip
475. Chevron
476. Chicken Wire
477. Chickling
478. Child of the Forest
479. Chilton Foliat
480. Chris Peer
481. Chris Sanders
482. Christiansburg
483. Christianus
484. Christine
485. Chris Lauten
486. Christmas Cheer
487. Christmas Wish

488. Chrome Yellow
489. Chthonic
490. Chubby
491. Chugayster
492. Cicely Hall
493. Cicely's Tubby
494. Cider with Rosie
495. Cikcakasti
496. Ciklamin
497. Cilcicus
498. Cinderdine
499. Cinderella
500. Cindy
501. Cirencester Old Spot
502. Clare Blakeway-Phillips
503. Claud Biddulph
504. Claudia
505. Clemens Heidger
506. Cliff Curtis
507. Clifton Hambden
508. Clovis
509. Clown
510. Clun
511. Clun Green
512. Clun Green Convolute
513. Clun Green Plicate
514. Clun Queen
515. Cockatoo
516. Col
517. Col. Leonid Bonderenko
518. Colesborne
519. Colesbourne Green Tip
520. Colossus
521. Comet
522. Comme if Taut
523. Compton Court
524. Compton Green Eye
525. Compu Ted
526. Condroz
527. Condroz Premature
528. Confetti
529. Conquest
530. Conundrum
531. Coolbalintaggart
532. Copek
533. Copton Trym
534. Cordelia
535. Corker
536. Corkscrew
537. Cornwood
538. Cornwood Gem
539. Corona North
540. Corrie McKeague
541. Corrin
542. Coton Manor
543. Cotswold Beauty
544. Cotswold Farm
545. Cottisford
546. Courteenhall
547. Covertside
548. Cowhouse Green
549. Craigton Twin
550. Creme Anglaise
551. Creole
552. Cressida
553. Crimea
554. Crimean Emerald
555. Crinkle Crankle
556. Crinoline
557. Crinolinum
558. Cristata
559. Cronkhill Green Tip
560. Cross Eyes
561. Cupid
562. Curly
563. Curry
564. Curtsey
565. Cutie Pie
566. Cyclops
567. Cyril Warr
568. Daglingworth
569. Daglingworth Kasaba
570. Daglingworth Yellow
571. Dainty Maid
572. Daisy Hynes
573. Daisy May
574. Daisy Sunshine
575. Dame Margot Fonteyn
576. Dangling Green
577. Danish Gold
578. Danish Swan
579. Danube Star
580. Danube Valley
581. Daphne's Maximus
582. Daphne's Scissors
583. Das Gelbe Vom Ei
584. Davey Nice
585. David
586. David Baker
587. David Bromley
588. David Bromley's Early
589. David Culp
590. David Quinton
591. David Shackleton
592. David's Seersucker
593. De Dreie Muskatiers
594. De Kanibaal
595. December
596. December Green Tip
597. Decima McAuley
598. Decora
599. D'Ecouvet Vert
600. Deeley Boppers
601. Deer Slot
602. Désse Verte
603. Demo
604. Denton
605. Der Fliegende Holländer
606. Der Grüne Fürst
607. Der Grüne Verzauberer
608. Derwish
609. Desdemona
610. Devon Marble
611. Diana Broughton
612. Dick's Early yellow
613. Dicke Tante
614. Dickerchen
615. Dickkopf

616. Diggory
617. Diggory Like
618. Ding Dong
619. Dionysus
620. Discovery
621. Distinction
622. Dizzy
623. Doddington
624. Dodo Norton
625. Don Armstrong
626. Don Hackonberry
627. Don Pedro
628. Donald Sims Early
629. Donaugold
630. Doncaster's Double Charmer
631. Doncaster's Double Scharlock
632. Donna Buang
633. Doornduyn Piet
634. Doornduyn Pippa
635. Doornduyn Siep
636. Dopey
637. Doppel Zappler
638. Dora Parker
639. Doris Page
640. Doris Potter
641. Dorothy Foreman
642. Dorothy Lucking
643. Dotty
644. Dot Underhill
645. Double Bill
646. Double Entendre
647. Double Green Tips
648. Double Top
649. Down and Out
650. Dr Tony Rodgerson
651. Dragonfly
652. Dragoon
653. Dream
654. Dreaming Spires
655. Drei Ohren
656. Dreispitz
657. Dreisporn
658. Dreycott Greentip
659. Drummonds Giant
660. Druon Antigoon
661. Dryad Artemis
662. Dryad Cerberus
663. Dryad Demeter
664. Dryad Echo
665. Dryad Gold
666. Dryad Gold Bullion
667. Dryad Gold Charm
668. Dryad Gold Ingot
669. Dryad Gold Medal
670. Dryad Gold Nugget
671. Dryad Gold Ribbon
672. Dryad Gold Sovereign
673. Dryad Gold Star
674. Dryad Hera
675. Dryad Venus
676. Dubbelduin
677. Duckie
678. Duet
679. Dumpy Denton
680. Dumpy Green
681. Dunley Hall
682. Dunskey Talia
683. Durris
684. Duxford Giant
685. Dwarf Green Danube
686. Dvajni Zajcek
687. Dvocvetna Krinolinco
688. Dvostebelni
689. Dymock
690. E. A. Bowles
691. Earl Green
692. Earliest
693. Earliest of All
694. Early Autumn Jewel
695. Early Riser
696. Early Start
697. Early to Rize
698. Early Twin
699. Echoes
700. Echoes of Christmas
701. Éclair
702. Ecusson d'Or
703. Eden
704. Edinburgh Ketton
705. Edith
706. Egret
707. Eifenlantaarntjes
708. Eilys Elizabeth Hartley
709. Ein Hauch von Lauch
710. Eisbar
711. Eiskristall
712. Eisternshen
713. Elcatus
714. Eleanor's Double
715. Eleanor Blakeway Phillips
716. Elegans
717. Elegant Bear
718. Eleni
719. Elexir
720. Elfenlantaarntjes
721. Elfin
722. Eliot Hodgkin
723. Elise Scharlock
724. Elizabeth Benfield
725. Elizabeth Harrison
726. Elizabeth Parker-Jervis
727. Ellen
728. Ellen Minnet
729. Elles Dream
730. Ellie Boardman
731. Ellie Saunders
732. Elmley Lovett
733. Else Bauer
734. Else Grollenberg
735. Elsje Mitchell
736. *elwesii*
737. *elwesii* Hiemalis Group
738. *elwesii* var. *elwesii*
739. *elwesii* var. *monostictus*
740. Elworthy Bumble Bee
741. Em
742. Emerald

743. Emerald Bear
744. Emerald Bells
745. Emerald Hughes
746. Emerald Isle
747. Emma Mackenzie
748. Emma Thick
749. Emmelina
750. Emperor Augustus
751. En Garde
752. Enid Bromley
753. Envy
754. Eppie
755. Epiphany
756. Eric E
757. Eric Fisher
758. Eric Watson
759. Eric's Choice
760. Ermine Ad Astra
761. Ermine Farm
762. Ermine Green
763. Ermine House
764. Ermine Joyce
765. Ermine Lace
766. Ermine Oddity
767. Ermine Ruby
768. Ermine Spiky
769. Ermine Street
770. Ermintrude
771. Erway
772. Erythrae
773. Eshley D
774. Eslington Halfer
775. Esmerelda
776. Essex Girl
777. Essie Huxley
778. Esther Merton
779. Ethiebeaton
780. Eutopia
781. Eva
782. Eva Turner
783. Evenley Double
784. Evienia
785. Ewe
786. Ex Broadleigh Gardens
787. Ex Cox
788. Ex Guy De Schryver
789. Ex Herbert Ransome
790. Ex Lowick Form
791. Ex Margaret Rivers
792. Ex Martin Rix
793. Ex Ruby Hall
794. Ex Warham
795. Excelsis
796. Eye Eye
797. Eye Shadow
798. Eyebright Early
799. Fabian
800. Faint Heart
801. Fair Maid
802. Fairlight
803. Fairy Tail
804. Fairy Tails Flair
805. Faith Stewart Liberty
806. Fake Pearls
807. Falkland House
808. Fanfare
809. Fanny
810. Fanfreluche
811. Faringdon Double
812. Farthinghoe Beauty
813. Fascination
814. Fat Boy
815. Fatty
816. Fatty Arbuckle
817. Fatty Puff
818. Fedegrüner
819. Federkleid
820. Federschwingen
821. Fee Clochette
822. Feintripp in Weiss
823. Fenella
824. Fenstead End
825. Feodora
826. Ferdinand von Rayski
827. Fieldgate A
828. Fieldgate Allegro
829. Fieldgate Continuo
830. Fieldgate E
831. Fieldgate Forte
832. Fieldgate Fortissimo
833. Fieldgate Fugue
834. Fieldgate Imp
835. Fieldgate L
836. Fieldgate Prelude
837. Fieldgate Primrose Legacy
838. Fieldgate Sophie
839. Fieldgate Superb
840. Fieldgate Tiffany
841. Fifi
842. Fifty Shades of Green
843. Filgree
844. Filou
845. Finchale Abbey
846. Finger in the Dyke
847. Fiona Mackenzie
848. Fiona's Gold
849. Fireworks
850. First Lady
851. Fishing Rod
852. Flamenco Dancer
853. Flanders
854. Flanders Beauty
855. Flanders Field Green
856. Flanders Field Yellow
857. Flanders Proud
858. Flattering
859. Flavescens
860. Flavus
861. Fliether Glocke
862. Flight of Fancy
863. Flocon de Neige
864. Flore Pleno - *elwesii*
865. Flore Pleno - *nivalis*
866. Florence Baker
867. Flotte Lotte
868. Fluff
869. Fly Away Peter
870. Fly Fishing
871. Fly Fly
872. Foresight
873. Forest Muse
874. Forger's Brush
875. Forge Double

876. Fortune
877. *fosteri*
878. *fosteri* var. *antepenensis*
879. Fotini
880. Foundling
881. Four By Four
882. Four English Ones
883. Four Four Time
884. Four Play
885. Foursome
886. Fox Farm
887. Foxgrove Magnet
888. Foxton
889. Framlingham Double
890. Francesca De Grammont
891. Frank Lebsa
892. Frank Turner
893. Franz Hadacek
894. Franz Joseph
895. Fraulein Else
896. Frazeri
897. Freda
898. Fred Fiegal Double
899. Fred's Double Scharlock
900. Fred's Giant
901. Fred's Surprise
902. Freddy King
903. Freddy's White Wonder
904. Friar Tuck
905. Fridge Magnet
906. Friedle
907. Frieve
908. Frilsham Spider
909. Frinton Advent
910. Frodo
911. Frohnauer Gold
912. Frühes Elegantes
913. Full Marks
914. Fun Ribbons
915. Funny Justine
916. Funny Peculiar
917. Fuzz
918. G 71
919. G 75
920. G. 77
921. G. F. Handel
922. Gabriel
923. Gaishaaugen
924. Galadriel
925. Galatea
926. Galaxy
927. Gandalf
928. GC 70
929. Gedenkemein
930. Geert Groote
931. Geisterschwingen
932. Gekke Henkie
933. Gelbe Marlu
934. Gelber Clown
935. Gemini
936. Genet's Giant
937. Genie
938. George Chiswell
939. George Chiswell No 1
940. George Chiswell No 7
941. George Chiswell No 9
942. George Elwes
943. George Proverbs
944. Georgia
945. Georgia Arula
946. Gerald Bauer
947. Gerard Parker
948. Gerlinda
949. Gestrichelt Pulk
950. Getatzt
951. Ghost
952. Ghost Spirit
953. Giant
954. Gill Gregory
955. Gills Special
956. Gimli
957. Gimley Din
958. Ginn's Byzantinus
959. Ginn's Imperati
960. Gipfelstürmer
961. Glacier
962. Gladysdale
963. Glass Marble
964. Glen Chantry Green Tip
965. Glenchantress
966. Glenorma
967. Globosus
968. Glooming
969. Gloria
970. Glory Gold
971. Gloriette
972. Gloucester Old Spot
973. Glücksglöckchen
974. Glowing
975. Goatee Green Tip
976. Goblet
977. Godfrey Owen
978. Gold Digger
979. Gold Double
980. Gold Dust
981. Gold Edge
982. Gold Fever
983. Gold Finger
984. Gold Shield
985. Gold Heart
986. Gold Vein
987. Goldcrest
988. Golden Boy
989. Golden Chalice
990. Golden Fleece
991. Golden Girl
992. Golden Glow
993. Golden Promise
994. Goldfinch
995. Goldmine
996. Gold 'n' Green
997. Goldrandröckcheng löckchen
998. Goliath
999. Gone Fishing
1000. Good Blue Leaved
1001. Gora
1002. Gottqaldii

1003. Goudgloren
1004. Government House
1005. Grace
1006. *gracilis*
1007. *gracilis* Virsecent Form
1008. Grake's Gold
1009. Grake's Green Bells
1010. Grake's Herediji
1011. Grake's Monster
1012. Grakes Oddity
1013. Grake's Silver Spoon
1014. Grake's Yellow
1015. Grande Billiet
1016. Grand Juhe
1017. Grandior
1018. Granny Smyth
1019. Grant Colvin
1020. Grasshopper
1021. Grave Concern
1022. Gravesend Giant
1023. Gravity
1024. Gray's Child
1025. Grayswood
1026. Green Arrow
1027. Green Bear
1028. Green Bottle
1029. Green Brush
1030. Green Cage
1031. Green Claw
1032. Green Comet
1033. Green Crown
1034. Green Dancer
1035. Green Day
1036. Green Desire
1037. Green Diamond
1038. Green Early Jewel
1039. Green Envy
1040. Green Epaulet
1041. Green Eyes
1042. Green Feathers
1043. Green Feelings
1044. Green Fingers
1045. Green Flake
1046. Green Flash
1047. Green Flight
1048. Green Fresh
1049. Green Gauge
1050. Green Genes
1051. Green Goblet
1052. Green Goblin Twin
1053. Green Grey and White
1054. Green Hayes
1055. Green Heik
1056. Green Hell
1057. Green Hornet
1058. Green Ibis
1059. Green Ice
1060. Green Jeanie
1061. Green Leaf
1062. Green Keeper
1063. Green Lantern
1064. Green Leaved Hybrid ex R. D. Nutt
1065. Green Light
1066. Green Maid
1067. Green Man
1068. Green Mark
1069. Green Mile
1070. Green Mist
1071. Green Necklace
1072. Green of Hearts
1073. Green Peace
1074. Green Peacock
1075. Green Pips
1076. Green Poc Shock
1077. Green Ribbon
1078. Green Shadow
1079. Green Spider
1080. Green Stripe
1081. Green Tear
1082. Green Teeth
1083. Green Testament
1084. Green Tipped *elwesii*
1085. Green Tipped
1086. Green Tipped Balloon
1087. Green Tipped Double
1088. Green Tips
1089. Green Tornadoes
1090. Green Tube
1091. Green Wasp
1092. Green Whisp
1093. Green Wings
1094. Green Woodpecker
1095. Green X
1096. Green Zebra
1097. Greenbottle
1098. Greenfield
1099. Greenfinch
1100. Greenfingers
1101. Greenish
1102. Greenkeeper
1103. Greenmantle
1104. Greenpeace
1105. Greenshank
1106. Greentips
1107. Greenzzly
1108. Grim Reaper
1109. Gruin
1110. Grumpy
1111. Grun Bis In Die Nagelspitzen
1112. Grün Fürst
1113. Grüne Donauzwerg
1114. Grüne Ostern
1115. Grüne Ostern 11
1116. Grüne Perle
1117. Grüne Schwerter
1118. Grüne Sptiz
1119. Grüne Tatzen
1120. Grüne Waldfee
1121. Grüne Weihnacht
1122. Grünspecht
1123. Grünspecht No 4
1124. Grüner Faun
1125. Grüner Fruhauf-steher
1126. Grüner Gnom
1127. Grüner Kakadu
1128. Grüner Langfinger
1129. Grüner Milan
1130. Grüner Nebel
1131. Grüner Pausback
1132. Grüner Pendelkugel

1133. Grüner Splitter
1134. Grüner Streifentropfen
1135. Grüner Winzling
1136. Grünes Gefieder
1137. Grünfrosch
1138. Grunschnabel
1139. Grünspecht
1140. Gryz
1141. Günter Bauer
1142. Günter Waldorf
1143. Günters Geist
1144. Gusmusi
1145. Gwen
1146. H. Purcell
1147. Haconby Early
1148. Haconby Green
1149. Haddon's Tiny
1150. Hagen Hastdunicht gesehn
1151. Hainborn
1152. Hainburger Poc
1153. Halcyon
1154. Half and Half
1155. Halfway
1156. Halo
1157. Hallelujah
1158. Halloween
1159. Hambutt's Brush
1160. Hambutt's Orchard
1161. Hanging Out
1162. Hannah Billiet
1163. Hanneke
1164. Hanning's Horror
1165. Hans Guck In Die Luft
1166. Hans Meyer
1167. Happy New Year
1168. Hardwick
1169. Harewood Twin
1170. Harlequin
1171. Haroula
1172. Hattrick
1173. Hawkeye
1174. Hawkshead
1175. Hazeldene
1176. Headbourne
1177. Heart's Desire
1178. Heaven's Patch
1179. Hedgehog
1180. Hedgehog Green Tip
1181. Heffalump
1182. Heidewitzka
1183. Helen
1184. Helen Tomlinson
1185. Helios
1186. Heloise des Essourts
1187. Henham No 1
1188. Henley Green Spot
1189. Henrietta
1190. Henry Broughton
1191. Henry's White Lady
1192. Hera
1193. Heracles
1194. Herbstflöckchen
1195. Hercule
1196. Here We Are
1197. Herr Specht
1198. Hessellund Autumn Snow
1199. Hidden Secret
1200. Hiemalis (ex Broadleigh Gardens)
1201. Highdown
1202. Highdown 457 Dwarf
1203. Hilary's Coquette
1204. Hildegard Owen
1205. Hill Poe
1206. Hill View
1207. Hippolyta
1208. Hobbgoblin
1209. Hõbevalge
1210. Hobsons Choice
1211. Hocus Pocus
1212. Hoddles Creek
1213. Hoggets Narrow
1214. Hoggets Round
1215. Holla Die Waldfee
1216. Hollis
1217. Holo Glob
1218. Hologram
1219. Hololeucus
1220. Holywell
1221. Homers Moly
1222. Homersfield
1223. Honeysuckle Cottage
1224. Honigmund
1225. Hornet
1226. Horst Kaufman
1227. Hortensis
1228. Hörup
1229. Hosentrager
1230. Hoverfly
1231. Howard Wheeler
1232. Howick Starlight
1233. Howick Yellow
1234. Hrastovlje
1235. Hugh Mackenzie
1236. Hulapalu
1237. Humpty Dumpty
1238. Hunton Early Bird
1239. Hunton Giant
1240. Hunton Herald
1241. Huttlestone
1242. Hyde Lodge
1243. I Knew It
1244. Ian Christie
1245. Icarus
1246. Icarus Spiky Double
1247. Ice Bear
1248. Ice Crystal
1249. Ice Princess
1250. Icicle
1251. Ida Maud
1252. Igraine
1253. *ikariae*
1254. *ikariae* 'Butts Form'
1255. Ilse Bilse
1256. Imbolc
1257. Imitatio
1258. Imperati
1259. Impudent
1260. In The Works
1261. Indegracht

1262. Indiskretion
1263. Ingrid
1264. Ingrid Bauer
1265. Ione Hecker
1266. Irish Green
1267. Isabell Blakeway-Phillips
1268. Iseghem
1269. Ismail
1270. Ispahan
1271. Istrian Cream
1272. Istrian Dwarf
1273. Ivy Cottage Corporal
1274. Ivy Cottage Green Tips
1275. J. Haydn
1276. Jabot
1277. Jack Hynes
1278. Jack In The Green
1279. Jack Mead
1280. Jack Percival
1281. Jacqueline
1282. Jade
1283. Jade Bear
1284. Jade Feathers
1285. Jade Joanne
1286. James Backhouse
1287. Jamie Broughton
1288. Jane Nicholls
1289. Janet
1290. Janet Aspland
1291. Janet's Gold
1292. Janet Cropley
1293. January Sales
1294. Janus
1295. Jaquenetta
1296. Jason Scharlok
1297. Jaspar
1298. Jean's Double
1299. Jedburgh
1300. Jennifer Hewitt
1301. Jennifer Rogers
1302. Jenny Owen
1303. Jenny Wren Hybrid
1304. Jenny Wren *nivalis*
1305. Jenny's Pearl
1306. Jess Egerton
1307. Jesse Jane
1308. Jessica
1309. Jimmy Platt
1310. JMDS-7B2-6
1311. Joan May
1312. Joan Weighell
1313. Joe's Poculiform
1314. Joe Sharman
1315. Joe's Spotted
1316. Joe's Yellow
1317. John Gray
1318. John Long
1319. John Marr
1320. John Nash
1321. John Owen
1322. John Tomlinson
1323. John's Y Fronts
1324. Jonathon
1325. Jos Mens
1326. Joshua Jansen
1327. Josie
1328. Joy Cozens
1329. Joyce
1330. JRM3139
1331. Jubilee Green
1332. Judith
1333. Judith's Yelbino
1334. Jule Jansen
1335. Julia
1336. Julie
1337. July
1338. June
1339. June Boardman
1340. Jupe's Bell
1341. Just The Two of us
1342. Just White
1343. Käina
1344. Kalum
1345. Kardamili
1346. Karla Tausendschön
1347. Karneval
1348. Kaspar In Grün
1349. Kastellorhizo
1350. Katarina
1351. Kath Dryden
1352. Katherine James
1353. Kathleen Beddington
1354. Katie Campbell
1355. Kato
1356. Katsoris
1357. Kedworth
1358. Keith Lamb
1359. Kelpie
1360. Kelways
1361. Kencot Ivy
1362. Kencott Kali
1363. Kencot Pickles
1364. Kencott Pip
1365. Kencott Ripple
1366. Kennemerend
1367. Kenneth Becket AM Form
1368. Kenneth Hall
1369. Kermit
1370. Kermode Bear
1371. Kersen
1372. Ketton
1373. Ketzkhovelii
1374. Kew
1375. Kew Form
1376. Kew Green
1377. Keyhole
1378. Kif
1379. Kija
1380. Kildare
1381. Kilkenny Giant
1382. Kilverstone
1383. Kin McIntosh
1384. Kinetic
1385. Kingennie
1386. King of Spades
1387. King of the Green Tipped Pocs
1388. Kingston Double
1389. Kinn Macintoch
1390. Kinnaird
1391. Kirsty
1392. Kirtling Tower

1393. Kite
1394. Klešče
1395. Knickerbockerrockchen
1396. *koenenianus*
1397. Koper-Capodistria
1398. Kosorog
1399. Krabat
1400. *krasnovii*
1401. *krinolinca*
1402. Krinolinca Quartet
1403. Kristin Meier
1404. Kryptonite
1405. Kubalach
1406. Kudos
1407. Kugelkompakt
1408. Kullaka
1409. Kurt Kleisa
1410. Kyre Park
1411. L. P. Long
1412. L. P. Short
1413. La Bohème
1414. La Morinière
1415. La Vie En Rose
1416. Lac de Balcère
1417. Lacewing
1418. Ladham's Variety
1419. Ladyboy
1420. Ladies Fingers
1421. Lady Ainsworth
1422. Lady Alice
1423. Lady Beatrix Stanley
1424. Lady Dalhousie
1425. Lady Elephinstone
1426. Lady Fairhaven
1427. Lady Latife
1428. Lady Lorna
1429. Lady Mary Grey
1430. Lady Moore
1431. Lady Putman
1432. Lady Scharlock
1433. Lady X
1434. Ladybird
1435. Lagle
1436. *lagodechianus*
1437. Lake Garda
1438. Lambrook Green
1439. Lambrook Greensleeves
1440. Lambrook Tall
1441. Lampijonček
1442. Lampshade
1443. Lanarth
1444. Lange Kerls
1445. Lange Wapper
1446. Langenweddinger Knuddel
1447. Lapwing
1448. Latest of All
1449. Lubfrosch
1450. Lavinia
1451. Lazybones
1452. Le Suiss Charlie
1453. Lea
1454. Leckford Form
1455. Lefki
1456. Legolas
1457. Leilani
1458. Lemon and Lime
1459. Lemon Drops
1460. Lemon Kindness
1461. Lemon Twist
1462. Lemongrass
1463. Leopard
1464. Lerinda
1465. Let It Be
1466. Lets Do It
1467. Leyla
1468. Liam Schofield
1469. Liberty
1470. Lichtgeel
1471. Lichtes Grün
1472. Liebsten
1473. Light Bulb
1474. Lili Vegrin
1475. Lime Trym
1476. Limetree
1477. Limey
1478. Lin
1479. Linientreu
1480. Linnett Green Tips
1481. Little Angie
1482. Little Artist
1483. Little Aud
1484. Little Beauty
1485. Little Bell
1486. Little Ben
1487. Little Big Horn
1488. Little Bitton
1489. Little Carol
1490. Little Dancer
1491. Little Dorrit
1492. Little Drip
1493. Little Emma
1494. Little Italy
1495. Little Green Grasshopper
1496. Little Joan
1497. Little John
1498. Little Late Comer
1499. Little Magnet
1500. Little Prince
1501. Little Star
1502. Livia
1503. Livia Augusta
1504. Ljubljana
1505. Llo 'n' Green
1506. Lobeliocvetni
1507. Lockers Delight
1508. Lodestar
1509. Lohengrin
1510. Long Drop
1511. Long Guy
1512. Long John
1513. Long John Silver
1514. Long Leys
1515. Long Peo
1516. Long Tall Sally
1517. Long Wasp
1518. Lonely Island
1519. Longbow
1520. Longfellow
1521. Longiflorus
1522. Longheart
1523. Longner Hall
1524. Longraigue

1525. Longstowe
1526. Longwood Green Tip
1527. Longworth Double
1528. Look At Me
1529. Look At Yourself
1530. Look Up Twin
1531. Looking Around
1532. Loose Spirit
1533. Lop Ears
1534. Lord Kitchener
1535. Lord Lieutenant
1536. Lord Monostictus
1537. Louise Ann Bromley
1538. Lovely Rika
1539. Lowick
1540. Luca
1541. Luca Cords
1542. Lucas William
1543. Lucifer
1544. Lucy
1545. Luke
1546. Lulu
1547. Lutea
1548. Lutescens
1549. Lutz Bauer
1550. Lydiard Diana
1551. Lyn
1552. Lynch Green
1553. Lyzzick
1554. MT4027
1555. Macedonicus
1556. Msculata
1557. Madame M
1558. Madelaine
1559. Mafangza
1560. Magda
1561. Magdalen Erskine
1562. Magdeburg Giant
1563. Magic
1564. Maggi in The Green
1565. Maggi Young
1566. Maggie In The Green
1567. Magnet
1568. Magnus
1569. Maid Marian
1570. Maidwell C
1571. Maidwell L
1572. Mainstream
1573. Majestic
1574. Major
1575. Major Pam
1576. Majus
1577. Maja's Gift
1578. Makellos
1579. Mala Kronica
1580. Malachite
1581. Malvose
1582. Mamma Mia
1583. Mandarin
1584. Manor Farm Early
1585. Mantis
1586. March Sunshine
1587. Marchwood
1588. Margaret Ann
1589. Margaret Biddulph
1590. Margaret Billington
1591. Margaret Ford
1592. Margaret Markham
1593. Margaret Mitchell
1594. Margaret Owen
1595. Margaret's Star
1596. Margery Fish
1597. Maria
1598. Marianne Hogener
1599. Marielle
1600. Marijke
1601. Marilyn Munroe
1602. Marin
1603. Marion's Magnet
1604. Marjorie Brown
1605. Mark Solomon
1606. Mark's Tall
1607. Market Cross
1608. Marlene Uhlhaas
1609. Marlie Raphael
1610. Marmin
1611. Marshall's Green
1612. Marmot Chartreuse
1613. Marmot Cross
1614. Marmot Duncan
1615. Marmot Flight
1616. Marmot Ghost
1617. Marmot Koksilah
1618. Marmot Puckered
1619. Marmot Shocker
1620. Marmot Stripe
1621. Marmot Two
1622. Martha Maclaren
1623. Martin
1624. Mary Ann Gibbs
1625. Mary Biddulph
1626. Mary Heley-Hutchin son
1627. Mary O'Brien
1628. Marzenbecher
1629. Märzun's Maigrün
1630. Maskenball
1631. Matoula
1632. Matt Bishop
1633. Matt-adors
1634. Maximus
1635. May Jupe
1636. Mayfair Chapel
1637. MBAC 93
1638. Medena
1639. Medno
1640. Megan
1641. Melanie Broughton
1642. Melanie S
1643. Melanie's Bloomers
1644. Melbourne
1645. Melfield
1646. Melnik
1647. Melody in Green
1648. Melot
1649. Melvillei
1650. Melvin Jope
1651. Melvyn's Hope
1652. Melvyn's Orange
1653. Mephisto
1654. Merladok
1655. Merlin
1656. Merlin Bonds Form
1657. Merlin's Autumn Child

1658. Mette
1659. Mex
1660. Michael Holecroft
1661. Michael Myers Greentipped
1662. Midas
1663. Midge
1664. Midori
1665. Midwinter
1666. Mighty Atom
1667. Mill House
1668. Mill View
1669. Miller's Late
1670. Milk and Honey
1671. Min
1672. Mini Me
1673. Miniehaha
1674. Miniskirt
1675. Minnie the Moucher
1676. Minnie Warren
1677. Miss Adventure
1678. Miss Behaving
1679. Miss Demeanor
1680. Miss Ellie
1681. Miss Hassell's Hybrid Cultivar
1682. Miss Mowcher
1683. Miss Prissy
1684. Miss Tep
1685. Miss Tilly
1686. Miss Willmott
1687. Missenden Splendour
1688. Mmm
1689. Moby Dick
1690. Moccas
1691. Modern Art
1692. Molly Watts
1693. Mona
1694. Money
1695. Monica
1696. Monk
1697. Monti Pincenti
1698. Moonlight
1699. Moonlight Shadow
1700. Moon Shadow
1701. Moonshine
1702. Moortown
1703. Moortown Mighty
1704. Moose Bear
1705. Morag
1706. Mordred
1707. Moreton Mill
1708. More Than a Feeling
1709. Morgana
1710. Moses Basket
1711. Mosquito
1712. Mostly Ghostly
1713. Motabile
1714. Mother Goose
1715. Mothering Sunday
1716. Mount Everest
1717. Mountain Sunrise
1718. Moya's Green
1719. Mr. Blobby
1720. Mr Frayling's Double
1721. Mr. Omer
1722. Mr. Paganini
1723. Mr. Peggotty
1724. Mr Positive
1725. Mr. Spoons
1726. Mr. Stinker
1727. Mr Taylor
1728. Mr Thompson
1729. Mr. Winkler
1730. Mrs Backhouse No 12
1731. Mrs H. Maxwell
1732. Mrs McNamara
1733. Mrs Norris
1734. Mrs Thompson
1735. Mrs Tiggywinkle
1736. Mrs W. M. George
1737. Mrs White
1738. Mrs Wrightson's Double
1739. MT4027
1740. Muku
1741. Mulberry
1742. Munchkin
1743. Mundy Newborough
1744. Mura
1745. Murska Sobota
1746. Music
1747. Must Have
1748. Mustachio
1749. Mustang Sally
1750. My Turn
1751. Myddleton Giant
1752. Mystra
1753. Nadelstreifenspitzen
1754. Naomi Slade
1755. Napoleon
1756. Narcisocvetni
1757. Narrengesicht
1758. Narwhal
1759. Natalie Garton
1760. Naughton
1761. Neat Green
1762. Neckless Wonder
1763. Neill Fraser
1764. Nell-e-Belle
1765. Nellie Brinsley
1766. Nelly's Birthday
1767. Nelly's Nets
1768. Nerissa
1769. Neon Kiss
1770. Nero
1771. Nessie
1772. Netherhall Double Yellow
1773. Netherhall Yellow
1774. Netopir
1775. Nettetaler Gold
1776. Never Before
1777. Newby
1778. Newry Giant
1779. Nicana
1780. Nice n Early
1781. Nigel Chadwick
1782. Nigel Colborn
1783. Nightlight
1784. Nightowl
1785. *nivalis*
1786. Noalene

1787. Nobody is Perfect
1788. Nobody's Prank
1789. Nona Goriana
1790. Norfolk Blonde
1791. Norfolk Small
1792. Norfolk Late
1793. Norris
1794. North Star
1795. North Star Bear
1796. Northern Lights
1797. Northgrange
1798. North Green Wasp
1799. Nothing Special
1800. Nova Amore
1801. Nova Gorica
1802. Nova Goriana
1803. Novelty
1804. November
1805. November Snow
1806. Novers Green
1807. Nutt's Double
1808. Nutt's Early
1809. Octavia Augustus
1810. Octopussy
1811. Odd Sharlock
1812. O Mahoney
1813. Oh My God
1814. Oirlicher Elfe
1815. Ojeh
1816. Olana
1817. Old Court
1818. Old January
1819. Olive
1820. Oliver Wyatt's Green
1821. Oliver Wyatt's Green Tip
1822. Oluna's Mother
1823. Omega
1824. One Drop or Two
1825. Onkle Oscar
1826. Oosterhouw
1827. Ophelia
1828. Oreanda
1829. Orion
1830. Orleton
1831. Orlicocvetni
1832. Orwell Greentip
1833. Ostergras
1834. Otto Fauser
1835. Over The Fence
1836. Paddy's Ketton
1837. Pagoda
1838. Pajec
1839. Palava
1840. Pale Cross
1841. Pale Face
1842. Pale Sue
1843. Pallidus
1844. Pan
1845. *panjutinii*
1846. Pannonia
1847. Papageno
1848. Paradise Giant
1849. Paraglider
1850. Paranoid
1851. Parcel
1852. Paris
1853. Pasjezobi
1854. Party Dress
1855. Pastures Green
1856. Pat Bender
1857. Pat Mackenzie
1858. Pat Mason
1859. Pat Schofield
1860. Patricia Ann
1861. Pausbäckchen
1862. Pausbacke
1863. Peardrop
1864. Pearl Drops
1865. Peg Sharples
1866. Pegasti Nos
1867. Pelican
1868. Pendulus
1869. Penelope Ann
1870. Penny
1871. Peppermint Candy
1872. Pepys
1873. Percy Picton
1874. Peregrin
1875. Perfect
1876. Perfection
1877. Peroxide Blonde
1878. Perrots Brook
1879. Persephone
1880. *peshmenii*
1881. Peter Fry
1882. Peter Gatehouse
1883. Peter Gooding
1884. Peter Joy
1885. Peter Pan
1886. Petite Belgiē
1887. Petite Fill de Flandres
1888. Petlistni
1889. Petticoat
1890. Pewsey Green
1891. Pewsey Vale
1892. Phaenika Samos
1893. Phantom
1894. Phantomas
1895. Phidias
1896. Phil Bryn
1897. Phil Cornish
1898. Phil's Fancy
1899. Philippe André Meyer
1900. Phoebe
1901. Phuk
1902. Pictus
1903. Pieces of Eight
1904. Pieter's early Giant
1905. Pingpoc
1906. Pink Panther
1907. Piping Plover
1908. Pisces
1909. Pixie
1910. *platyphyllus*
1911. Platytepalus
1912. Plemy Green
1913. Plenissimus
1914. *plicatus*
1915. *plicatus* subs *byzantinus*
1916. *plicatus* subs. *plicatus*
1917. Plover

1918. Plummet
1919. Plymouth Rock
1920. Pocahontas
1921. Poculiformis
1922. Poculi-Perfect
1923. Podsreda
1924. Poesie Der Provinz
1925. Poison
1926. Polar Bear
1927. Polar Breeze
1928. Pom-Pom
1929. Poor Boy
1930. Popcorn
1931. Popov
1932. Porcelain Charm
1933. Porlock No 2
1934. Poseidon
1935. Potter's Prelude
1936. Poweltown
1937. Praecox
1938. Prague Spring
1939. Praha
1940. Prestwood White
1941. Pretty Close
1942. Pride O' The Mill
1943. Primrose Hill
1944. Primrose Hill Special
1945. Primrose's Giant
1946. Primrose Warburg
1947. Princeps
1948. Pringle
1949. Prinzenrock
1950. Prior Park Hybrid
1951. Priscilla Bacon
1952. Pritlikavec
1953. Progaste Krinolince
1954. Proliferation
1955. Propellerköpfchen
1956. Puck
1957. Puffball
1958. Puffin
1959. Pumilis
1960. Pummelchen
1961. Pumpkin
1962. Pumpot
1963. Punctatus
1964. Pusey Green Tips
1965. Pyramid
1966. Quad
1967. Quadriga
1968. Quadrimont
1969. Quasimodo
1970. Quatrefoil
1971. Quintet
1972. Rabbit Ears
1973. Rachelae
1974. Radar Love
1975. Ragamuffin
1976. Ragini
1977. Rainbow Alchemist
1978. Rainbow Eyes of March
1979. Rainbow Farm Early
1980. Rainbow Five Star
1981. Rainbow Melchior
1982. Rainbow Pinky and Perky
1983. Rainbow Richard ll
1984. Rainbow Snowgoose
1985. Rainbow Triceratops
1986. Rainbow Trio
1987. Rainbow Yo Yo
1988. Rainbows End
1989. Ramsay
1990. Ransom
1991. Ransom's Dwarf
1992. Ransom's Late
1993. Rapture
1994. Ratz Fatz
1995. Raveningham
1996. Ray Cobb
1997. Raysky Double
1998. Rebecca
1999. Red House
2000. Redoutei
2001. Reflexus
2002. Reflection
2003. *reginae-olgae*
2004. *reginae-olgae* subs *reginae-olgae*
2005. *reginae-olgae* subsp *vernalis*
2006. Remember Remem ber
2007. Rev. Hailstone
2008. Rheingold
2009. Richard Ayres
2010. Richard Blakway-Phillips
2011. Richard Hand scombe
2012. Richard Nutt
2013. Richard Todd
2014. Rings Rum
2015. Risky
2016. Rita Rutherford
2017. *rizehensis*
2018. Rob
2019. Robert Berkeley
2020. Robert Wijnen
2021. Robin
2022. Robin Hall
2023. Robin Hood
2024. Robustus
2025. Robyn Janey
2026. Rocket Ship
2027. Rodmarton
2028. Rodmarton Aldebaran
2029. Rodmarton Arcturus
2030. Rodmarton Capella
2031. Rodmarton Regulus
2032. Rodmarton Sirius
2033. Roger's Rough
2034. Romeo
2035. Ron Ginns
2036. Ronald Frank
2037. Ronald Mackenzie
2038. Rose Baron
2039. Rose Lloyd
2040. Rosemary Burnham
2041. Rosemary Mitchell
2042. Rosie
2043. Rosie Dear
2044. Rosie Larkin
2045. Roudnice

2046. Roulade
2047. Roundhead
2048. Route 66
2049. Rowallane
2050. Rowan Russell
2051. Ruben
2052. Ruby Baker
2053. Ruby's Gebertstag-blumen
2054. Ruby's Gold
2055. Ruby's Green Dream
2056. Ruby's Surprise
2057. Ruhr Valley
2058. Rumenoglavi
2059. Rumpelstiltskin
2060. Rupert Golby
2061. Rus Ukraine
2062. Ruslan
2063. Rusch'l Jo
2064. Rushmere Green
2065. Ruth
2066. Ruth Birchall
2067. Ruth Dashwood
2068. Ruth MacLaren
2069. Ryton Ruth
2070. Ryzzards Grun
2071. SA0901
2072. S. Arnott
2073. Sabine
2074. Sally Ann
2075. Sally Pasmore
2076. Sally Wickenden
2077. Sally's Double
2078. Samantha
2079. *samothracicus*
2080. Samwise
2081. Sandersii
2082. Sandhill Gate
2083. Santa Claus
2084. Sara
2085. Saraband
2086. Sarah Dumont
2087. Sarah's Sweet Heart
2088. Satelite
2089. Savill Gold
2090. Savrasska
2091. Schalschwinger
2092. Scharlockii
2093. Scharlockii JM3
2094. Scharlockii JM8
2095. Scharlockii Trifth
2096. Scharloqueen
2097. Schattenwinkel
2098. Schloss Wiedenfeld
2099. Schloss Wolfsgarten
2100. Schlyters Dwarg
2101. Schmalhans Im Grünen
2102. Schneeweisschen
2103. Schnuckelchen
2104. Schorbuser Blut
2105. Schorbuser Irrlicht
2106. Schorbuser Lampion
2107. Schwanenweiss
2108. Schwefelgeschwafel
2109. Scissors
2110. Schloss Wolfsgarten
2111. Scone Palace
2112. Scrooge
2113. Seagull
2114. Seersucker
2115. Selbourne Green Tips
2116. Selina Cords
2117. Senna Sunrise
2118. Sentinel
2119. Sentvid
2120. Seppel
2121. September
2122. Septet
2123. Seraph
2124. Serotinus
2125. Seremony
2126. Sextet
2127. Sgt Major
2128. Shades of Green
2129. Shadow
2130. Shaggy
2131. Shampita
2132. Sharman's Late
2133. Sheds and Outhouses
2134. Sheila MacQueen
2135. Shellbrook Giant
2136. Shepherd's Crook
2137. Shepton Merlin
2138. Shimmer
2139. Shirley Cross
2140. Shoij
2141. Shooting Star
2142. Shrek
2143. Shrimp
2144. Shropshire Queen
2145. Shrubbery Special
2146. Siamese Triplet
2147. Siamese Twin
2148. Siamski Cetvorčki
2149. Siamski Dvojčki
2150. Siamski Trojčki
2151. Sibbertoft Magnet
2152. Sibbertoft Manor
2153. Sibbertoft No 2
2154. Sibbertoft Soldier
2155. Sibbertoft White
2156. Sickle
2157. Silverwells
2158. Silvia
2159. Simone
2160. Simon Lockyer
2161. Simplex
2162. Simply Glowing
2163. Sioux
2164. Sir Edward Elgar
2165. Sir Henry B. C.
2166. Sir Herbert Maxwell
2167. Sir Paul Dewilde
2168. Siskin
2169. Six Leaves
2170. Sky Rocket
2171. Sixtus
2172. Skyward
2173. Slim Jim
2174. Smagg
2175. Smaragdi
2176. Smaragdsplitter
2177. Smiley
2178. Smokey Mountain

2179. Smoking Joe
2180. Smurf
2181. Snocus
2182. Snoopy
2183. Snow Angel
2184. Snow Clock
2185. Snow Flurry
2186. Snow Fox
2187. Snow Queen
2188. Snow White
2189. Snow White Gnome
2190. Snow White Ex Hulsman
2191. Snow White's Gnome
2192. Snowball
2193. Sofia
2194. Solo
2195. Son of a Gun
2196. Sophie North
2197. Sortež
2198. Sotor
2199. South Esk
2200. South Hayes
2201. Sova
2202. Spanish Swan
2203. Sparkler
2204. Spartacus
2205. Spetchley
2206. Spetchley Cassaba
2207. Spetchley Yellow
2208. Spili
2209. Spindlestone Surprise
2210. Splendid Cornelia
2211. Spring Coat
2212. Spring Greens
2213. Springs Gold
2214. Spring Pearl
2215. Spot
2216. Springvale
2217. Springwood
2218. Springwood Park
2219. Spirit Bear
2220. Sprite
2221. Spröder Adel
2222. Squib
2223. Squidly
2224. Squire Burroughs
2225. Srček
2226. St Andrews Cross
2227. St Annes
2228. St Boniface
2229. St Pancras
2230. St Patrick's Day
2231. St Sylvestre
2232. Standing Tall
2233. Star Bear
2234. Starbright
2235. Starburst
2236. Starling
2237. Stavroula
2238. Stefan
2239. Stenopetalus
2240. Sterretjes
2241. Steveni
2242. Steve's Yellow
2243. Stork
2244. Straffan
2245. Strasbourg Twins
2246. Streifengelaut
2247. Streifentroll
2248. Streifen-Weh
2249. Stromae
2250. Stuyvensberg
2251. Sulphur Ramblings
2252. Styx
2253. Such A Beauty
2254. Sündhaft Grün
2255. Sunlime
2256. Sunrise
2257. Sunny Bunny
2258. Super Magnet
2259. Susan Grimshaw
2260. Sutton Court
2261. Sutton Courtenay
2262. Swallow
2263. Swan Lake
2264. Swanton
2265. Sweet Alice
2266. Sweet Gaia
2267. Sweetheart
2268. Sweety
2269. Sweta
2270. Sybil Roberta
2271. Sybil Stern
2272. Sylvie
2273. Tall Dark and Handsome
2274. Tall Prague Spring
2275. Tänte Kathe
2276. Tantrum
2277. Tatiana
2278. Tatjana
2279. Taubenei
2280. Tauricus
2281. Taurus
2282. Tausendschön
2283. Taylor's Giant
2284. Taynton Squash
2285. Teardrop
2286. Teresa
2287. Terry Jones
2288. Tessera
2289. The Apothecary
2290. The Big Bopper
2291. The Bogeyman
2292. The Bowl
2293. The Bride
2294. The Charmer
2295. The Dragon
2296. The Eyes Have It
2297. The Face
2298. The Flying Dutchman
2299. The Green Bucket
2300. The Groom
2301. The Hulk
2302. The Iliad
2303. The Lady with the Lamp
2304. The Linns
2305. The More The Merrier
2306. The Musketeer
2307. The Naked Truth
2308. The Paddle
2309. The Pearl
2310. The Reaper

2311. The Superbowl
2312. The System
2313. The Tower
2314. The Whins Yellow
2315. The Whopper
2316. The Wizard
2317. Theresa Stone
2318. Thin Green
2319. Three Leaves
2320. Three Ships
2321. Thumbelina
2322. Tilebarn Jamie
2323. Till E
2324. Till Sonnenschein
2325. Tilly
2326. Tim's Hill Poe
2327. Timber Wolf
2328. Timpany Bols
2329. Timpany Dwarf
2330. Timpany Late
2331. Tinkers House Double
2332. Tinus
2333. Tinus Snor
2334. Tiny
2335. Tiny Sunshine Glow
2336. Tiny Tim
2337. Tiny Whitey
2338. Tippy Green
2339. Tipsy
2340. Titania
2341. Titanic
2342. Toby
2343. Todered
2344. Tom Watkins
2345. Tomišelj
2346. Tommy
2347. Tomoko
2348. Tomtit
2349. Tootsie
2350. Tom's Fool
2351. Topaz
2352. Toy Soldier
2353. Tramlines
2354. Tramplet
2355. *transcaucasicus*
2356. Traumänzer
2357. Treasure Island
2358. Tricia Tay
2359. Tricky Treasure
2360. Tricorn
2361. Trifolius
2362. Trimmer
2363. Trinity
2364. Trois Jardins
2365. Trojan
2366. *trojanus*
2367. Trotter's Giant Plicate
2368. Trotters Merlin
2369. Trudy
2370. Truffle
2371. Trumpolute
2372. Trumps
2373. Trumpses
2374. Trust
2375. Trym
2376. Trym Baby
2377. Trym Ingram
2378. Trymest
2379. Trymlet
2380. Trymming
2381. Trymposter
2382. Tryzm
2383. Tsoe-Tsoe Mama
2384. Tubby Merlin
2385. Tuesday's Child
2386. Tuff
2387. Turban der Scherifen
2388. Turncoat
2389. Tusker
2390. Tutu
2391. Twice As Grumpy
2392. Twilight
2393. Twin Star
2394. Twinkletoes
2395. Twins
2396. Twister
2397. Two Eyes
2398. Two Pair
2399. Two Spots
2400. Twosome
2401. Unbrensis
2402. Unbricus
2403. Uli's Gift
2404. Uncle Dick
2405. Undecover
2406. Under Cherry Plum
2407. Underdog
2408. Unguiculatus
2409. Upcher
2410. Uptown Girl
2411. Uranium
2412. Utopia
2413. Utrecht Goud
2414. Utrecht Lime
2415. Valentine
2416. Valentine's Day
2417. Valentine's Gift
2418. Valérie
2419. Valerie Finnis
2420. Vammen
2421. Vamp
2422. Van Houttei
2423. Vasiliki
2424. Vegetarians
2425. Veliki Progasti
2426. Vera Trum
2427. Verdigris
2428. Verdure
2429. Veronica Cross
2430. Vertigo
2431. Veržej
2432. Vic Horton
2433. Victor
2434. Victoria Gibbs
2435. Viervierteltakt
2436. Vijolicovetni
2437. Vile
2438. Vipava
2439. Virescens
2440. Virgin
2441. Viridans
2442. Viridapice
2443. Vivere
2444. Vlaase Reus
2445. Vladimir Vasak

2446. Vornehma Blāsse
2447. Vrammen
2448. Vlam Re
2449. Vrtoče
2450. W. A. Mozart
2451. W O Backhouse
2452. Waffles
2453. Wake Up Call
2454. Walker Canada
2455. Walloping Whopper
2456. Walrus
2457. Walter Fish
2458. Waltham Place 2
2459. Waltham Place 3
2460. Wandin
2461. Wandlebury Ring
2462. Wanted
2463. Warande's Groenpunt
2464. Warande's Grootste
2465. Warande's Sieraad
2466. Warburg No 1
2467. Warburg No 2
2468. Warburg No 5
2469. Warburton
2470. Wardance
2471. Warei
2472. Warham
2473. Warham Rectory
2474. Warley Belles
2475. Warley Duo
2476. Warley Longbow
2477. Warley Place
2478. Warmes Licht
2479. Warwickshire Gemini
2480. Warwickshire Gremlin
2481. Washfield Colesbourne
2482. Washfield Merlin
2483. Washfield Titania
2484. Washfield Warham
2485. Wasp
2486. Watchmeister Dimpfelmoser
2487. Watson L
2488. Waverley Aristocrat
2489. Waverley Irish Colleen
2490. Waverley Little Egret
2491. Wayside
2492. Weather Hill
2493. Webb's Gold
2494. Wedding Dress
2495. Wee Betty
2496. Wee Grumpy
2497. Weisse Adventsperlen
2498. Weisse Miriam
2499. Weisse Pendelkugel
2500. Weisses Gold
2501. Weisses Lauten
2502. Wellenspiel
2503. Welsh Whiskers
2504. Welshway
2505. Wendover Green Tipped
2506. Wendy's Gold
2507. Werner Reinerman
2508. Wessex Ploughman
2509. Wessex Titan
2510. West Porlock No 8
2511. Westburn
2512. Whirligig
2513. Whisky
2514. Whisper
2515. Whitalli Foxgrove Form
2516. White Admiral
2517. White Cloud
2518. White Day Dream
2519. White Dream
2520. White Gem
2521. White Kay
2522. White Light
2523. White Madness
2524. White Magic
2525. White Mischief
2526. White Perfection
2527. White Stag
2528. White Sugar
2529. White Swan
2530. White Wings
2531. Whiteheart
2532. Whiter Than White
2533. Whittalli Foxgrove Form
2534. Whittington
2535. Wifi Bingo
2536. Wifi Magic Mint
2537. Wifi Eden's Basket
2538. Wifi Scha Nef
2539. Wifi Tobias
2540. Wifi Walk the Line
2541. Wilhelm Bauer
2542. William Louis
2543. William Thomson
2544. Willie The Weeper
2545. Wim
2546. Wind Turbine
2547. Windesheim
2548. Windmill
2549. Windsor Yellow
2550. Winifrede Mathias
2551. Winklepicker
2552. Winner
2553. Winsley Hall
2554. Winter Green
2555. Winter Memory
2556. Winterwood
2557. Wishbone
2558. Wisley Magnet
2559. Wisp
2560. Wispy
2561. Witchwood
2562. Witton
2563. Wodney Muž
2564. Wol Staines
2565. Wolfgang
2566. Wolfgang's Gold
2567. Wommelghem
2568. Wonder Green
2569. Wonston Double
2570. Woodfield
2571. Woodpeckers
2572. Woodtown
2573. Woodwire

2574. Woottens Limey
2575. Woozle
2576. *woronowii*
2577. Wraysbury
2578. Wrentnall Peculiar
2579. xalleni
2580. X-Files
2581. xhybridus
2582. Xmas
2583. xvalentinei
2584. XXX
2585. Yaffle
2586. Yamanlar
2587. Yarnton
2588. Yashmak
2589. Yellow Angel
2590. Yellow Sparkle
2591. Yes We Can
2592. Yeti
2593. York Minster
2594. Yuletide
2595. Yvonne
2596. Yvonne Hay
2597. YYY
2598. Zabika
2599. Zah Zuh Zaz
2600. Zajček
2601. Zarnikova
2602. Zarte Versuchung
2603. Zauberfee
2604. Zelen
2605. Zeleni A
2606. Zeleni Pacek
2607. Zhooshi Starburst
2608. Zlatko
2609. Zminec
2610. Zone Bear
2611. Zorro
2612. Zuckerpuppe
2613. Zvezdica
2614. Zwanenberg

Extinct or Previously Used *Galanthus* Names

Extinct or Previously Used *Galanthus* Names

The following names have been in circulation in the past but the snowdrops are either now extinct or have been re-named. These names therefore should not be used again.

Acme
Aestivalis
Aestivum
Afterglow
Aidin
Albus
Allen's Perfection
Allen's Seedling
Anacreon
Anamus
Angustifolius
Aurora

Bacherweise
Balloon
Beauty
Belated
Benthall Beauty
Biflorus
Billinghurst Lime
Blonde Inga
Biscapus
Boydii
Boyd's Green
Boyd's Green Double
Breviflos

Cambo Estate Form
Candidus
Canon Ellacombe
Carpenter's Shop
Cassaba Boydii
Cassalia
Cathcarti
Cathcartiae
Chapeli
Charmer
Charmer Flore Pleno
Claudia
Cloven
Clown
Clun Green
Colesbourne
Colesbourne Colossus
Colesbourne Seedling
Cool Ballintaggert
Coquette
corcyrensis
Cottage Corporal
Creole
Cronkhill Selection
Cryptonite
Cupid
Curio

Delight
Demo
Dionysius
Distinction
Dora Parker
Double Charmer
Double Green
Dragoon
Drumont Giant
Dunrobin Seedling

Early Bird
Early To Rise
Edmonsham Emma
Eleana Blakwey-Phillips
Elegans
Elegenas
Elise Scharlock
Elsae
Emerald
Empress
Ermine Shuttlecock
Ermine Windmill
Erythrae
Esmerelda
Excelsior

Faldonside
Fascination
Fenella
Fieldgate B
Finale
Flavescens
Flavescens Flore Pleno
Flavus
Flore Pleno Green Tip
Fortune
Francesca
Frazeri

Gem
Gimli
Globosus
Gottwaldii
Grandiflorus
Grandior
Grandis
Grayling
Green Horror
Green Lantern
Green Tip Double
Greener
Greenfield
Grenadier
Gusmusi

Hagen's No 1
Harold Wheeler
Helen Dillon's Whopper
Hololeucus

Howick
Hughes Emerald
Humberts Orchard
Hybridus

Ivy Cottage No 6

Janis Rucksans
Jason Scharlock
Jenny

Kedworth
Kew
Kilkenny Giant
King

Ladham's Variety
Lady Norris
Lammermuir SnowWhite
Largest of All
Lazybones
Leonid Bondarenko
Leopard
Limetrees
Loiterer
Longiflorus
Lop Ears
Lutescens Flore Pleno
Lyminghame

Macedonicus
Maculatus
Majestic
Major
Major Pam's Double
Majus
Marshall's Green
Marvel
Maximus
Meg
Megan Morris
Melvillei Major
Mikwood
Mill House
Minnie Warren
Miriam Ledger
Miss Demeanor
Miss Hassells Hybrid Cult.
Missendon Narrow
M. Myers Green Tipped
Mocca's
Mrs Backhouse
Mrs Backhouse's Spectacles
Mystra

Nancy Lindsay
Newry Giant
Nicana
Nip
No Mark
Norm's Late
North Green Only
Novelty
November
November Merlin

Ochrospelius
Octobrensis
Olivia
O Mahoney
Oliver Wyatt
Oliver Wyatt's Giant
Omega

Pallidus
Pamski's Double
Paris
Pat Schofield
Pearl
Pendulus
Perfection
Phuk
Phaenika Samos
Picton's Mighty Atom
Pictus
Platytepalus
Plenissimus
Plover
Poculiformis
Praecox
Pumilis
Punctatus

Quaderipetala

Rabbit Ears
Rachelae
Raphael
Rebecca
Redoutei
Reflexus
Rita
Rita Rutherford
Robert Berkeley
Robin
Robustus
Romeo
Ruksans

Samuel Arnott
Sandersi
Saville Yellow
Scharlockii Flore Pleno
Serotinus
Shackleton
Sibbertoft
Silvia
Snogerupii
Spindlestone Yellow
Spot
Stargazer
Stenopetalus
Steveni
Sutton Supreme

Talisman
Tauricus
Taurus
The Charmer
The O'Mahoney
The Stalker
The Tower
Tinkerbell
Titania (single)
Tomtit
Trifolius

Umbrensis
Umbricus

Unguiculatus

Valentine
Valentinae
Valentine
Valentinei
Van Houttei
Victor
Victoria Gibbs
Virgin
Viridans
Viridapices
Virid-apice

W. Thomson
Warham
Warham Variety
West Porlock No 1
West Porlock No 7/9
White Lady
Whittalilii
William Ball No 1
Winner
Wraysbury

Yellow Trym
Yorkshire Cottage

CITES

In the early 1990s, the collection of wild snowdrops came under control of CITES – the Convention on International Trade in Endangered Species of Wild Fauna and Flora.

This is an agreement between governments world wide to help ensure the survival of endangered species.

In most countries it is now illegal to collect snowdrops without a CITES permit, whether bulbs or living or dead plants. It does however allow limited trade in three species, G. nivalis, G. elwesii and G. woronowii from Turkey and Georgia.

All snowdrops are listed under CITES Appendix 2.

Wildlife & Countryside Act 1981

It is illegal to collect any bulbs from the wild in the UK without the landowner's permission. This is covered by the 1981 Wild Life and Countryside Act in the UK, and in Northern Ireland the 1985 Wildlife (Northern Ireland) Order.

Under the Theft Act 1968 it is a serious offence to remove plants from the countryside for commercial purposes without formal authorization. Severe penalties are imposed by the UK National Wild Crime Unit and police.

Where to See and Buy Snowdrops

United Kingdom

Andrew's Snowdrops, Andrew and Anita Thorp, Bungalow No 5, Main Street, Theddingworth, Near Lutterworth, Leicestershire, LE17 6QZ.

Ashwood Nurseries, Ashwood Lower Lane, Ashwood, Kingswinford, West Midlands, DY6 0AE. www.ashwoodsnurseries.com

Avon Bulbs, Burnt House Farm, Mid-Lambrook, South Petherton, Somerset, TA13 5HE.
www.avonbulbs.com

Beth Chatto Gardens and Nursery, Elmstead Market, Colchester, Essex, CO7 7DB.
www.bethchattoshop.co.uk

Broadleigh Gardens, Barr House, Bishops Hall, Taunton, Somerset, TA4 1AE. www.broadleighbulbs.co.uk

Cambo Estate, Kingsbarns, St. Andrews, Scotland, KY16 8QD.
www.cambosnowdrops.com

Colesbourne Gardens LLP, Estate Office, Colesbourne, Cheltenham, Gloucestershire, GL53 9NP.
www.colesbournegardens.co.uk

Edrom Nurseries, Coldingham, Eyemouth, Berwickshire, Scotland, TD14 5TZ.
www.edrom-nurseries.co.uk

Edulis Nursery, Paul Barney, The Walled Garden, Tidmarsh Lane, Pangbourne, Berksire RG8 8HT.
Email: edulisnursery@gmail.com

Elworthy Cottage Plants, Mike and Jenny Spiller,
Elworthy, Taunton, Somerset, TA4 3PX.
www.elworthy-cottage.co.uk

Evenly Wood Gardens, Evenley, Northamptonshire, NN13 5SH.
www.evenleywoodgarden.co.uk

Field of Blooms, Ballymackey, Nenagh, Co. Tipperary, Ireland.
www.fieldofblooms.com

Foxgrove Plants, Foxgrove Farm, Skinners Green, Enborne,
Newbury, RG14 6RE.
www.foxgroveplants.co.uk

Glen Chantry, Wickham Bishop, Witham, Essex, CM8 3LG.
www.glenchantry.demon.co.uk

Harvey's Garden Plants, Great Green, Thurston, Bury St Edmunds,
Suffolk, IP31 3SJ.
www.harveysgardenplants.co.uk/galanthus.asp

Judy's Snowdrops, 19 Rugby Road, Long Lawford, Rugby,
Warwickshire, CV23 9DS.
www.judyssnowdrops.co.uk (Mail order only)

Monksilver Nursery, Oakington Road, Cottenham, Cambridge, CB24 8TW.
www.monksilvernursery.co.uk

Morlas Plants (formerly Cornovian Snowdrops) Woodlands, Nant Lane, Selattyn, Oswestry, Shropshire SY10 7HA.
Email: jane@morlasplants.co.uk
www.morlasplants.co.uk

North Green Snowdrops, North Green Only, Stoven, Beccles, NR34 8DG.

Pottertons Nursery, Moortown Road, Nettleton, Caistor, Lincolnshire, LN7 6HX.
www.pottertons.co.uk

Rainbow Farm, Michael and Anne Broadhurst, Halesworth Road, Reddisham, Beccles, NR34 8NE. Email broadhurst320@btinternet.com

Netherlands

De Boschhoeve Garden and Nursery, Boschoeve 3, 6874 NB Wolfheze, www.boschhoeve.nl

De Warande, Jan de Jager Road 2, AN 6998, Low Keppel.
www.dewarande.nl

The Hidden City Garden, Tuinontwerp, Groenadvies, Assendorperstraat, 178 8012 CE Zwolle.
www.tuinharrypierik.nl

Belgium

Coolplants, Cathy Portier, Margareta van Vlaanderenstraat 27, 8310 Bruges, Belgium.
Tel +32 50 68 43 58.
Email info@coolplants.com

Koen van Poucke, Heistraat 106, B 9110 Sint Niklaas. www.koenvanpouke.be

Kalmthout Arboretum and Botanical Garden, Heuval 2, Kalmthout, B-2920, www.arboretumkalmthout.be
Naturalised snowdrops, annual 'Snowdrop Show'.

Ruben Billiet, Meibosstraat 25, 8780, Oostrozebeke, Belgium.

Germany

Garten in den Wiessen, Hagen Engelmann, Torgauer-strasse 11, 03 048 Cottibus.
www.engelmannii.de
Garden openings, specialist snowdrop collection, sales.

Oirlicher Blumengarten, Oirlich 9, 41334
Nettetal, Hinsbeck.
Special snowdrop event, mail order sales.
www.oirlicher-blumengarten.de

USA

Brandywine Snowdrops, Brandywine Cottage, Down-ingtown, Pennsylvania.
www.davidlculp.com

Brent and Becky's Bulbs
www.brentandbeckysbulbs.com

Carolyn's Shade Gardens, 325 South Roberts Road, Bryn Mawr, PA 19010.
www.carolynsshadegardens.com
Email carolyn@carolynsshadegardens.com
Mail order, local sales and events.

Edelweiss Perennials Inc, 29800 S Barlow Road, Canby, Oregon 97013

Edgewood Gardens, Dr John Lonsdale, Exton Pennsyilvania
Email: info@edgewoodgardens.net
www.Edgwoodgardens.net

Ernie Cavallo Dormant Snowdrop list usually available first week in July.
Email: ernestcavallo@aol.com

Far Reaches Farm, 1818 Hastings Avenue. Port Town-send, Washington 98368
www.farreachesfarm.com

McClure and Zimmerman
www.mzbulb.com

Old House Gardens -
www.oldhousegardens.com

Temple Nursery, Hitch Lyman, PO Box 591, Trumansburg, NY 14886.
Snowdrop events and sales.

Terra Ceia Farms
www.terraceiafarms.com

Van Engelen Bulbs
www.vanengelen.com

Odyssey Bulbs, Russell Stafford, 43 Nisbet Street, Providence
Rhode Island 02906
Email: mail@odysseybulbs.com

Paula Squitiere's Snowdrops - Private message Paula on Facebook.

Australia

Sylvan Vale, Olinda Creek Road, Kalorama, VIC 3766.
www.svnltd.com

Merry Garth, Davies Lane, Mount Wilson, NSW 2786.

New Zealand

Jury Gardens, 589 Otaraoa Road, Waitara 4383.
www.jury.co.nz
Displays of snowdrops July-August, gardens open daily August-March.

Maple Glen Gardens and Nursery, Wyndham Letterbox Road, Wyndham.
www.mapleglen.co.nz
Naturalised snowdrops July-August, snowdrop sales. Open all year.

Plant Heritage Collection Holders - *Galanthus*

Mrs and Mrs D. MacLennan, 14 St Georges Crescent, Carlisle, CA3 9NL
Email. byndes2@btinternet.com
This important collection was awarded scientific status in 2016.

Lady Caroline Erskine, Cambo Estate, Kingsbarns, St Andrews, Scotland KY16 8QD.
www.camboestate.com

Mr S. Owen, 127 Stoke Road, Linslade, Leighton Buzzard, LU7 25R

Glossary and Abbreviations

Aberrant – Malformed.
Acuminate - Tapering to a point.
Acute – Terminating in a sharp point.
AGM – Royal Horticultural Society Award of Garden Merit awarded to plants with outstanding qualities in the garden.
AM – Royal Horticultural Society Award of Merit for worthy plants exhibited at shows.
Albino – All white, with rounded flower shape.
Alpine – Plant native to the Alpine zone, suitable for growing in rock garden conditions.
Anthers – Part of stamen containing pollen.
Apex – Tip of a plant formation as in leaf, root or perianth segment.
Applanate – Vernation of leaf where two leaves are held flat against each other

Base – The lowest point of a structure according to its place of insertion.
Basal plate – Base of bulb from which roots grow.
Binomial – Two-part naming system employed in Linnaean plant classification. First part denoting genus, second part denoting species.
Bulb – Underground storage organ composed of fleshy leaf bases enclosing the growing point.
Bulbil – Immature bulb formed at base of parent bulb.
Bulblet – Small bulb or 'offset' growing from parent bulb.

Capsule – Dry or semi-fleshy fruit containing seeds.
Classification – Arrangement of plants and other organisms into groups. ie. Species grouped into genera, then families etc.
Clone – Identical plants obtained from a single parent by vegetative propagation.
Cross – Hybridization of plant by transferring pollen from anthers of one plant to stigma of another.
Cucullate – Hooded as in leaf or spathe.
Cultivar – Variant of a plant selected for its merits in cultivation. Cultivars can also be selected directly from wild populations
Cuneate – Triangular or wedge-shaped.

Distribution – Geographical areas in which wild plants grow.
Dormant – Applied to a resting plant usually when it makes little or no growth.
Double-flowered - Flowers with additional segments

Elaiosome – Small appendage on seed with substances attractive to insects aiding distribution.
Explicative – Vernation of leaf showing margins sharply rolled under.

f. – Abbreviation of forma.
Filament – Stalk of stamen which carries the anthers.
Forma – Botanical rank for a minor variation from the norm that is usually rather rare and does not usually reproduce as a population. eg. Double flowers.

Galanthophile – Term applied to person with a passion for snowdrops.
Genus – Botanical classification category containing allied species. May also be sub-divided into further sections within the genera.
Glaucescent – Leaves and stems with green-grey colouring and light waxy bloom.
Glaucous – Leaves and stems with silvery-grey-blue colouring and light waxy bloom.
Globular – Rounded, spherical, ball-like.
Goffering – Applied by Galanthophiles to the light crimping on edges at base of segments.
Green-tipped - Snowdrop flowers having green marks on the outer segments, usually towards the apex.

Habitat – A plant's natural environment.
Hybrid – New plant produced by crossing two distinct species or subspecies.

ICBN – International Code of Botanical Nomenclature covering plant naming procedures.
ICNCP – International Code of Nomenclature for Cultivated Plants covering naming of cultivated plants.
Inflorescence – Flowering parts of plant.
Introduction – A plant occurring in the wild through human intervention but also used in a horticultural sense to mean brought into cultivation.
Inverse Poculiform – When outer segments replicate inner.

Longitudinal – Running lengthways.

Matte – Surface dull, not shiny.
Microclimate – Area with a climate differing from the normal prevailing conditions.
Midrib – Centre vein of leaf.
Morphology – Appertaining to physical characteristics of plant or organism.

Native – Plant or organism that occurs specifically in a particular region or country.
Naturalised – Introduced species which has successfully adapted and reproduces in its new situation.
Naturalising – Bulbs and other plants grown in conditions simulating their natural habitats.
Nectar – Sweet substance usually found in flowers which attracts insects to aid pollination.
Nectary – Gland secreting nectar.
Notch – Small indentation at apex of segment also referred to as sinus.

Obtuse – Blunt or rounded applied to apex.

Ovary – Female part of flower which produces seeds after fertilisation.
Ovules – Contained in the ovary.

Pedicel – Stalk of individual flower.
Perianth – Term covering sepals and petals when they are indistinguishable from each other.
Perianth segment – Individual section of perianth.
Petal – Modified leaf form, often coloured, surrounding and protecting stamens and attracting insects for pollination.
Petaloid – Usually stamens or sepals taking on the appearance of petals.
Pistil – Entire female organs of flower.
Poculiform – Cup-shaped flower formed when segments are all of equal length.
Pollen – Male plant cells contained in anthers or pollen sacs.
Pollination – Transfer of pollen grains onto female stigma either naturally by wind or insects, or artificially by hand.
Propagation –Method of increasing plants either by seed or vegetative processes.
Prostrate – Lying flat to ground.

Reflexed – Usually referring to leaves or petals sharply bent back on themselves.

Scape – Single leafless flower stem particularly in bulbs.
Seedling – Plants raised from seed, or a young single unbranched stemmed plant after germination.
Segment – When petal and sepal cannot be distinguished from one another (perianth segment).
Sinus – Small notch at apex of segment.
Spathe – Structure composed of modified leaf or leaves enclosing bud or flower.
Species – Classification term applied to a single or group of closely related plants in a genus which breed true to type from seed.
Spiky – With reference to snowdrops, a flower with narrow segments that point upwards or nearly so rather than being pendant
Spp – Species.
Stamen – Complete male reproductive unit of flower containing filament and two anther lobes with pollen grains.
Sterile – Plants which do not or rarely set seed. Many double flowers are sterile as reproductive organs have become petals.
Stigma – Tip of female reproductive organ which secretes sticky solution prior to pollination.
Style – Stalk linking ovary and stigma in female flower.
Supervolute – Vernation with one leaf fully or partially wrapped around the other
Subsp. – Subspecies
Subspecies – Nomenclature group immediately below species, for a population of wild plants differing in some definable way from the norm.
Synonym – Alternative plant name where plant has been renamed or reclassified. Oldest name always takes priority if validly published.

Taxonomy – Study of plant classification.
Tepal – Individual perianth segment that cannot be differentiated into petal or sepal.

Tunic – Papery outer layer of compressed leaf scales surrounding bulb. New scales are continually formed inside bulb pushing older scales gradually outwards.

Unguiculate – Narrowed at base into claw shape.

Var. – Variety.
Variety – Next rank below sub-species in plant classification for plants showing minor but consistent variations from the norm.
Vernation – Arrangement of leaves in bud.
Virescent – Green stained segments.

Bibliography

Bishop, M., Davis, A. P. and John Grimshaw. *Snowdrops: A Monograph of Cultivated Galanthus*. Griffin Press, 2006.

Brickell, Christopher. *Royal Horticultural Society Encyclopaedia of Plants and Flowers*. Dorling Kindersley 2006.

Brickell, Christopher. *Royal Horticultural Society Encyclopaedia of Garden Plants*. Dorling Kindersley, 2003.

Cox,Freda. *A Gardener's Guide to Snowdrops*, (First and second editions) The Crowood Press.

Davis, A. P. 1999. *The Genus Galanthus*. Royal Botanic Gardens, Kew in association with Timber Press, Oregon.

Diduch, J. P. Red *Book of Ukraine*: Plant Kingdom. Globalkonsalting, Kyiv.

Griffiths, Mark. 1997. *The New RHS Dictionary Index of Garden Plants*. Macmillan.

IUCN *Red List of Threatened Species* (2011)

Phillips, Roger, and Martin Rix. *Bulbs*, Pan Books, 1989.

Sharman, Joe. Galanthus Gala Transcripts 1997-2011.

Tan, Kit; Burkhard Biel, and Sonja Siljak-Yakovlev. *Galanthus Samothracicus (Amaryllidaceae)* from the island of Samothraki, northeastern Greece. 2014.

Van Dijk, Hanneke. *Galanthomania*, Terra, 2011.

Waldorf, Günter. Schneeglökchen (Snowdrops - White Magic), August 2011.

Zobov, Dimitriy A, and Aaron Davis. *Galanthus panjutinii*. Magnolia Press, 2012.

Daffodil, Snowdrop and Tulip Year Book available from the Royal Horticultural Society.
RHS Database. www.rhs.org.uk

Other books by Freda Cox

Seasons In A Country Garden. Fulcrum Publishing USA, 1995

Designing and Creating a Mediterranean Garden. The Crowood Press, Wiltshire, 2005

Garden Style: An Essential Guide.The Crowood Press, Wiltshire, 2010

A Gardener's Guide to Snowdrops. The Crowood Press, Wiltshire, 2013

A Gardener's Guide to Snowdrops - Second Edition. The Crowood Press, Wiltshire, 2019.
www.crowood.com

Your Notes

www.ingramcontent.com/pod-product-compliance
Lightning Source LLC
Chambersburg PA
CBHW042000030726
47636CB00009B/500

* 9 7 8 1 8 4 3 9 6 6 9 1 3 *